AF658772

Particle Physics

Particle Physics

Yavari M. Nejad

Oxford Book Company
Jaipur ▪ New Delhi

Published By

Oxford Book Company

Regd. Off. "Bony Residency", Gate No. 2
Opp. Tilak Public School, Vishweswariya Nagar,
Gopalpura Road, Jaipur - 302018, India
Ph: 0141 - 2761280, Telefax: 2761381
email - oxfordbookjpr@gmail.com
website - oxfordbookcompany.com

Delhi Office: 102, 1st Floor, Satyam House,
4327/3, Ansari Road, Darya Ganj,
New Delhi - 110002, India
Ph: 011-45652440

Distributed By
KEYA INTERNATIONAL PRESS LLP
E-mail: keyainternationalpressllp@gmail.com

Published in India, 2024

© All Rights Reserved

ISBN : 9789350303917

All Rights Reserved. No part of this publication may be reproduced, stored in a retrieval system, or transmitted, in any form or by any means, electronic, mechanical, photocopying, recording, scanning or otherwise, without the prior written permission of the copyright owner.

Typeset by Shivangi Computers, Jaipur

Printed and bound in India by Thomson Press Ltd., New Delhi

Contents

Preface

Particle physics is a branch of physics that studies the elementary constituents of matter and radiation, and the interactions between them. It is also called "high energy physics", because many elementary particles do not occur under normal circumstances in nature, but can be created and detected during energetic collisions of other particles, as is done in particle accelerators. Modern particle physics research is focused on subatomic particles, which have less structure than atoms. These include atomic constituents such as electrons, protons, and neutrons (protons and neutrons are actually composite particles, made up of quarks), particles produced by radiative and scattering processes, such as photons, neutrinos, and muons, as well as a wide range of exotic particles. Strictly speaking, the term particle is a misnomer because the dynamics of particle physics are governed by quantum mechanics. As such, they exhibit wave-particle duality, displaying particle-like behaviour under certain experimental conditions and wave-like behaviour in others. All the particles and their interactions observed to date can be described by a quantum field theory called the Standard Model. The Standard Model has 40 species of elementary particles (24 fermions, 12 vector bosons, and 4 scalars), which can combine to form composite particles, accounting for the hundreds of other species of particles discovered since the 1960s. Experiments at particle accelerators, such as LEP, where sub-atomic particles collide at very high energies, reveal details of particlesand conditions that prevailed just after the Big Bang over 15 billion years ago.

Most experiments involve large international collaborations and are performed at overseas laboratories such as CERN in Geneva and DESY in Hamburg. These collaborations typically involve more than 300 people and the work at CERN is supported by 19 European countries. The accelerator is the basic tool of particle physics. It allows us to create the particle collisions that we want to study in our own laboratories. The high-energy collisions between particles that physicists are interested in do occurnaturally but the events are unpredictable and the number that can be observed (in cosmic rays) is low. Accelerators work by accelerating charged particles using electric fields. A linear accelerator accelerates particles in a straight line: the biggest linear machine, in Stanford, California, is two miles long. Circular machines are more common. As well as accelerating the particles using an electric field, circular accelerators bend their paths using a magnetic field. In a machine like LEP at CERN, where they have opposite charges, the particles being accelerated travel in opposite directions until they are forced to collide. The drawback is that the faster a particle travels, the harder it is to keep it moving in a circle but, in the largest circles (LEP is the largest in the world with circumference of 27km) less energy is wasted when accelerating particles to high speeds. Detectors are used to examine tracks made by the new particles that are produced when accelerated particles collide. In the early days photographic film, spark chambers and bubble chambers were used. Antimatter is very much like ordinary matter, but it carries the opposite charge. An anti-electron (a positively charged electron) is just another way of describing a positron. Crashing matter and antimatter together is now a daily occurrence in machines like LEP. The fact that the universe seems to be full of matter and not antimatter is one of the most baffling problems in modern physics. At the time of the Big Bang, matter and antimatter are believed to have been produced in equal quantities. What seems to have happened is that, at a somewhat later time, collisions between the two types have destroyed all the antimatter but left a little of the matter behind, from which our universe is made. The reason may be due to a tiny asymmetry in the way particles of matter and antimatter decay, thereby creating an excess of matter. The Big Bang theory correctly predicts that about 75% of all visible matter is hydrogen and about 25% helium. (All other matter accounts f or less than 1%.)Another great success of the theory is the presence of background microwave radiation in our universe, a relic of the Big Bang. There are many ideas about what dark matter might be, ranging from exotic new particles to black holes.

This comprehensive textbook integrates recent experimental results, including discovery of the Higgs boson, to convey the excitement of the field to graduate and postgraduate students.

—Yavari M. Nejad

Chapter 1

Introduction

Particle physics is a branch of physics that studies the existence and interactions of particles that are the constituents of what is usually referred to as *matter* or *radiation*. In current understanding, particles are excitations of quantum fields and interact following their dynamics. Most of the interest in this area is in *fundamental fields*, each of which cannot be described as a bound state of other fields. The current set of fundamental fields and their dynamics are summarized in a theory called the Standard Model, therefore particle physics is largely the study of the Standard Model's particle content and its possible extensions.

Figure: *Large Hadron Collider tunnel at CERN*

Subatomic Particles

Modern particle physics research is focused on subatomic particles, including atomic constituents such as electrons, protons, and neutrons (protons and neutrons are composite particles called baryons, made of quarks), produced by radioactive and scattering processes, such as photons, neutrinos, and muons, as well as a wide range of exotic particles. To be specific, the term *particle* is a misnomer from classical physics because the dynamics of particle physics are governed by quantum mechanics. As such, they exhibit wave-particle duality, displaying particle-like behaviour under certain experimental conditions and wave-like behaviour in others. In more technical terms, they are described by quantum state vectors in a Hilbert space, which is also treated in quantum field theory. Following the convention of particle physicists, *elementary particles* refer to objects such as electrons and photons as it is well known that these types of particles display wave-like properties as well. All particles and their interactions observed to date can be described almost entirely by a quantum field theory called the Standard Model. The Standard Model has 17 species of elementary particles: 12 fermions or 24 if distinguishing antiparticles, 4 vector bosons (5 with antiparticles), and 1 scalar boson. These elementary particles can combine to form composite particles, accounting for the hundreds of other species of particles discovered since the 1960s. The Standard Model has been found to agree with almost all the experimental tests conducted to date.

However, most particle physicists believe that it is an incomplete description of nature, and that a more fundamental theory awaits discovery. In recent years, measurements of neutrino mass have provided the first experimental deviations from the Standard Model.

Particle physics has affected the philosophy of science greatly. Some particle physicists adhere to reductionism, a point of view that has been criticized and defended by philosophers and scientists. Other physicists may defend the philosophy of holism, which has commonly been viewed to be reductionism's opposite.

History

The idea that all matter is composed of elementary particles dates to at least the 6th century BC. The philosophical doctrine of atomism and the nature of elementary particles were studied by ancient Greek philosophers such as Leucippus, Democritus, and Epicurus; ancient Indian philosophers such as Kanada, Dignâga, and Dharmakirti; medieval scientists such as Alhazen, Avicenna, and Algazel; and early modern European physicists such as Pierre Gassendi, Robert Boyle, and Isaac Newton. The particle theory of light was also proposed by Alhazen, Avicenna, Gassendi, and Newton. These early ideas were founded in abstract, philosophical reasoning rather than experimentation and empirical observation. In the 19th century, John Dalton, through his work on stoichiometry, concluded that each element of nature was composed of a single, unique type of particle. Dalton and his contemporaries believed these were the fundamental particles of nature and thus named them atoms, after the Greek word *atomos*, meaning "indivisible". However, near the end of the century, physicists discovered that atoms are not, in fact, the fundamental particles of nature, but conglomerates of even smaller particles.

The early 20th-century explorations of nuclear physics and quantum physics culminated in proofs of nuclear fission in 1939 by Lise Meitner (based on experiments by Otto Hahn), and nuclear fusion by Hans Bethe in the same year. These discoveries gave rise to an active industry of generating one atom from another, even rendering possible (although not profitable) the transmutation of lead into gold (alchemy). They also led to the development of nuclear weapons. Throughout the 1950s and 1960s, a bewildering variety of particles were found in scattering experiments. This was referred to as the "particle zoo". This term was deprecated after the formulation of the Standard Model during the 1970s in which the large number of particles was explained as combinations of a (relatively) small number of fundamental particles.

Standard Model

The very current state of the classification of elementary particles is the Standard Model. It describes the strong, weak, and electromagnetic fundamental interactions, using mediating gauge bosons. The species of gauge bosons are the gluons, W–, W+ and Z bosons, and the photons. The model also contains 24 fundamental particles, (12 particles and their associated anti-particles), which are the constituents of matter. Finally, it predicts the existence of a type of boson known as the Higgs boson, which is yet to be discovered.

Experimental Laboratories

In particle physics, the major international laboratories are:

- Brookhaven National Laboratory (Long Island, United States). Its main facility is the Relativistic Heavy Ion Collider (RHIC), which collides heavy ions such as gold ions and polarized protons. It is the world's first heavy ion collider, and the world's only polarized proton collider.

- Budker Institute of Nuclear Physics (Novosibirsk, Russia). Its main projects are now the electron-positron colliders VEPP-2000, operated since 2006, and VEPP-4, started experiments in 1994. Earlier facilities include the first electron-electron beam-beam collider VEP-1, which conducted experiments from 1964 to 1968; the electron-positron colliders VEPP-2, operated from 1965 to 1974; and its successor VEPP-2M, performed experiments in 1974-2000.
- CERN, (Franco-Swiss border, near Geneva). Its main project is now the Large Hadron Collider (LHC), which had its first beam circulation on 10 September 2008, and is now the world's most energetic collider of protons. It will also be the most energetic collider of heavy ions when it begins colliding lead ions in 2010. Earlier facilities include the Large Electron–Positron Collider (LEP), which was stopped in 2001 and then dismantled to give way for LHC; and the Super Proton Synchrotron, which is being reused as a pre-accelerator for LHC.
- DESY (Hamburg, Germany). Its main facility is the Hadron Elektron Ring Anlage (HERA), which collides electrons and positrons with protons.
- Fermilab, (Batavia, United States). Its main facility is the Tevatron, which collides protons and antiprotons and was the highest-energy particle collider in the world until the Large Hadron Collider surpassed it on 29 November 2009.
- KEK, (Tsukuba, Japan). It is the home of a number of experiments such as the K2K experiment, a neutrino oscillation experiment and Belle, an experiment measuring the CP violation of B mesons.

Many other particle accelerators exist: The techniques required to do modern experimental particle physics are quite varied and complex, constituting a sub-speciality nearly completely distinct from the theoretical side of the field.

Theory

Theoretical particle physics attempts to develop the models, theoretical framework, and mathematical tools to understand current experiments and make predictions for future experiments. There are several major interrelated efforts in theoretical particle physics today. One important branch attempts to better understand the Standard Model and its tests. By extracting the parameters of the Standard Model from experiments with less uncertainty, this work probes the limits of the Standard Model and therefore expands our understanding of nature's building blocks. These efforts are made challenging by the difficulty of calculating quantities in quantum chromodynamics. Some theorists working in this area refer to themselves as phenomenologists and may use the tools of quantum field theory and effective field theory. Others make use of lattice field theory and call themselves *lattice theorists.*

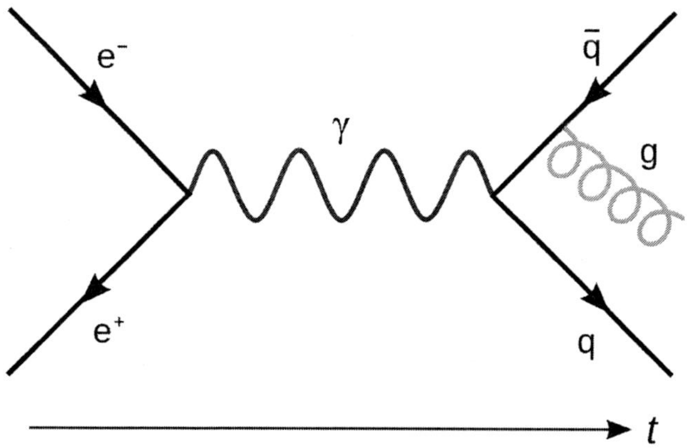

Figure: *Feynman diagram*

Another major effort is in model building where model builders develop ideas for what physics may lie beyond the Standard Model (at higher energies or smaller distances). This work is often motivated by the hierarchy problem and is constrained by existing experimental data. It may involve work on supersymmetry, alternatives to the Higgs mechanism, extra spatial dimensions (such as the Randall-Sundrum models), Preon theory, combinations of these, or other ideas.

A third major effort in theoretical particle physics is string theory. *String theorists* attempt to construct a unified description of quantum mechanics and general relativity by building a theory based on small strings, and branes rather than particles. If the theory is successful, it may be considered a "Theory of Everything". There are also other areas of work in theoretical particle physics ranging from particle cosmology to loop quantum gravity. This division of efforts in particle physics is reflected in the names of categories on the arXiv, a preprint archive: hep-th (theory), hep-ph (phenomenology), hep-ex (experiments), hep-lat (lattice gauge theory).

Practical Applications

As generations build upon others, potential applications increase in the use of particle physics technology. In 1930, the first hand-held cyclotron was built at Berkeley, California by Ernest O. Lawrence. More powerful accelerators were built soon after. The Berkeley cyclotron was later used to produce medical isotopes for research and treatment. The first application of this technology in the treatment of cancer was by Lawrence himself with his own mother as a patient. Medical science now uses particle beams in life saving technologies. This technology is also used in the superconducting of wires and cables. This is used for Magnetic Resonance Imaging magnets and ultimately the World Wide Web. Less known uses also include behavioural study of fluids and motions. Additional applications are found in medicine, homeland security, industry, computing, science, and workforce development illustrate a long and growing list of beneficial practical applications with contributions from particle physics.

Future

The overarching goal, which is pursued in several distinct ways, is to find and understand what physics may lie beyond the standard model. There are several powerful experimental reasons to expect new physics, including dark matter and neutrino mass. There are also theoretical hints that this new physics should be found at accessible energy scales. Furthermore, there may be unexpected and unpredicted surprises that will give us the most opportunity to learn about nature. Much of the efforts to find this new physics are focused on new collider experiments. The Large Hadron Collider (LHC) was completed in 2008 to help continue the search for the Higgs boson, supersymmetric particles, and other new physics. An intermediate goal is the construction of the International Linear Collider (ILC), which will complement the LHC by allowing more precise measurements of the properties of newly found particles. In August 2004, a decision for the technology of the ILC was taken but the site has still to be agreed upon.

In addition, there are important non-collider experiments that also attempt to find and understand physics beyond the Standard Model. One important non-collider effort is the determination of the neutrino masses, since these masses may arise from neutrinos mixing with very heavy particles. In addition, cosmological observations provide many useful constraints on the dark matter, although it may be impossible to determine the exact nature of the dark matter without the colliders. Finally, lower bounds on the very long lifetime of the proton put constraints on Grand Unified Theories at energy scales much higher than collider experiments will be able to probe any time soon.

Mechanical Wave

A mechanical, also called material wave, is a wave that needs a medium to travel. Mechanical waves cause a local oscillation of material. The oscillating material does not move far from its initial equilibrium position. Mechanical waves transport energy only. This energy propagates in the same direction as the wave. Any kind of wave (mechanical or electromagnetic) has a certain energy. No material is transported as a result of mechanical waves. A mechanical wave requires an initial energy input. Once this initial energy is added, the wave travels through the medium until all its energy is transferred. In contrast, electromagnetic waves require no medium, but can still travel through one.

Figure: *Ripple in water is a surface wave.*

One important property of mechanical waves is that their amplitudes possess an unusual form, displacement divided by reduced wavelength. When this gets comparable to unity, significant nonlinear effects such as harmonic generation may occur, and, if large enough, may result in chaotic effects. For example, waves on the surface of a body of water break when this dimensionless amplitude exceeds 1, resulting in a foam on the surface and turbulent mixing. The three types of mechanical waves are: transverse waves, longitudinal waves, and surface waves.

Types

Transverse Wave

Transverse waves cause the medium to vibrate at a right angle to the direction of the wave. Transverse waves have two parts—the crest and the trough. The crest is the highest point of the wave and the trough is the lowest. The wavelength is the distance from crest to crest or from trough to trough. Light has properties of a transverse wave, although it is an electromagnetic wave.

Longitudinal Wave

The speed of the longitudinal wave is increased in higher index of refraction, due to the closer proximity of the atoms in the medium that is being compressed. Sound is considered a longitudinal wave.

Surface Waves

This type of wave travels along a surface that is between two media. An example of a surface wave would be waves in a pool, or in an ocean, lake, or any other type of water body.

Examples

- Water
- Sound
- Ocean waves

Wave Equation

The wave equation is an important second-order linear partial differential equation for the description of waves – as they occur in physics – such as sound waves, light waves and water waves. It arises in fields like acoustics, electromagnetics, and fluid dynamics. Historically, the problem of a vibrating string such as that of a musical instrument was studied by Jean le Rond d'Alembert, Leonhard Euler, Daniel Bernoulli, and Joseph-Louis Lagrange. Wave equations are examples of hyperbolic partial differential equations, but there are many variations.

In its simplest form, the wave equation concerns a time variable t, one or more spatial variables $x_1, x_2, \ldots, x_n$, and a scalar function $u = u(x_1, x_2, \ldots, x_n; t)$, whose values could model the height of a wave. The wave equation for u is

$$\frac{\partial^2 u}{\partial t^2} = c^2 \nabla^2 u$$

where ∇^2 is the (spatial) Laplacian and where c is a fixed constant.

Solutions of this equation that are initially zero outside some restricted region propagate out from the region at a fixed speed in all spatial directions, as do physical waves from a localized disturbance; the constant c is identified with the propagation speed of the wave. This equation is linear, as the sum of any two solutions is again a solution: in physics this property is called the superposition principle. The equation alone does not specify a solution; a unique solution is usually obtained by setting a problem with further conditions, such as initial conditions, which prescribe the value and velocity of the wave. Another important class of problems specifies boundary conditions, for which the solutions represent standing waves, or harmonics, analogous to the harmonics of musical instruments. To model dispersive wave phenomena, those in which the speed of wave propagation varies with the frequency of the wave, the constant c is replaced by the phase velocity:

$$v_p = \frac{\omega}{k}.$$

Phenomena in which the speed depends on the amplitude of the wave are modelled by nonlinear wave equations:

$$\frac{\partial^2 u}{\partial t^2} = c(u)^2 \nabla^2 u$$

A wave may be superimposed onto another movement (for instance sound propagation in a moving medium like a gas flow). In that case the scalar u will contain a Mach factor (which is positive for the wave moving along the flow and negative for the reflected wave).

The elastic wave equation in three dimensions describes the propagation of waves in an isotropic homogeneous elastic medium. Most solid materials are elastic, so this equation describes such phenomena as seismic waves in the Earth and ultrasonic waves used to detect flaws in materials. While linear, this equation has a more complex form than the equations given above, as it must account for both longitudinal and transverse motion:

$$\rho \ddot{\mathbf{u}} = \mathbf{f} + (\lambda + 2\mu)\nabla(\nabla \cdot \mathbf{u}) - \mu \nabla \times (\nabla \times \mathbf{u})$$

where:

- λ and μ are the so-called Lamé parameters describing the elastic properties of the medium,
- ρ is the density,
- $\mathbf{f}$ is the source function (driving force),
- and $\mathbf{u}$ is the displacement vector.

Note that in this equation, both force and displacement are vector quantities. Thus, this equation is sometimes known as the vector wave equation. Variations of the wave equation are also found in quantum mechanics, plasma physics and general relativity.

Scalar Wave Equation in One Space Dimension

Derivation of the Wave Equation

From Hooke's law: The wave equation in the one dimensional case can be derived from Hooke's law in the following way: Imagine an array of little weights of mass m interconnected with massless springs of length h. The springs have a spring constant of k:

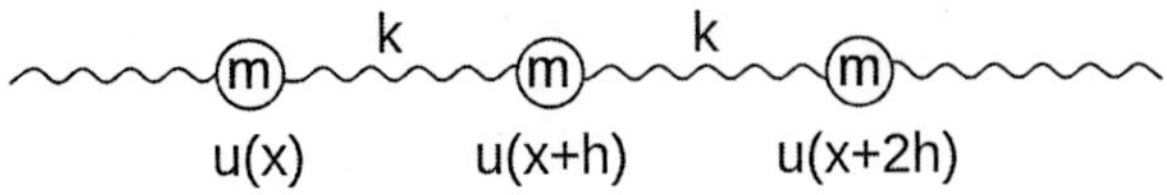

Here $u(x)$ measures the distance from the equilibrium of the mass situated at x. The forces exerted on the mass m at the location $x+h$ are:

$$F_{Newton} = m{\cdot}a(t) = m{\cdot}\frac{\partial^2}{\partial t^2}u(x+h,t)$$

$$F_{Hooke} = F_{x+2h} + F_x = k\left[u(x+2h,t) - u(x+h,t)\right]$$
$$+k[u(x,t) - u(x+h,t)]$$

The equation of motion for the weight at the location $x+h$ is given by equating these two forces:

$$m\frac{\partial^2 u(x+h,t)}{\partial t^2} = k[u(x+2h,t) - u(x+h,t) - u(x+h,t) + u(x,t)]$$

where the time-dependence of $u(x)$ has been made explicit.

If the array of weights consists of N weights spaced evenly over the length $L = N\,h$ of total mass $M = N\,m$, and the total spring constant of the array $K = k/N$ we can write the above equation as:

$$\frac{\partial^2 u(x+h,t)}{\partial t^2} = \frac{KL^2}{M}\frac{[u(x+2h,t) - 2u(x+h,t) + u(x,t)]}{h^2}$$

Taking the limit $N \to \infty, h \to 0$ (and assuming smoothness) one gets:

$$\frac{\partial^2 u(x,t)}{\partial t^2} = \frac{KL^2}{M}\frac{\partial^2 u(x,t)}{\partial x^2}$$

$(KL^2)/M$ is the square of the propagation speed in this particular case.

General Solution

The one dimensional wave equation is unusual for a partial differential equation in that a relatively simple general solution may be found. Defining new variables:

$$\xi = x - ct \quad ; \quad \eta = x + ct$$

changes the wave equation into

$$\frac{\partial^2 u}{\partial \xi \partial \eta} = 0$$

which leads to the general solution

$$u(\xi, \eta) = F(\xi) + G(\eta) \quad \Rightarrow \quad u(x,t) = F(x-ct) + G(x+ct)$$

In other words, solutions of the 1D wave equation are sums of a right travelling function F and a left travelling function G. "Travelling" means that the shape of these individual arbitrary functions with respect to x stays constant, however the functions are translated left and right with time at the speed c. This was derived by Jean le Rond d'Alembert.

Another way to arrive at this result is to note that the wave equation may be "factored":

$$\left[\frac{\partial}{\partial t} - c\frac{\partial}{\partial x}\right]\left[\frac{\partial}{\partial t} + c\frac{\partial}{\partial x}\right]u = 0$$

$$\Downarrow$$

$$\frac{\partial u}{\partial t} - c\frac{\partial u}{\partial x} = 0 \quad \text{and} \quad \frac{\partial u}{\partial t} + c\frac{\partial u}{\partial x} = 0$$

These last two equations are advection equations, one left travelling and one right, both with constant speed c. For an initial value problem, the arbitrary functions F and G can be determined to satisfy initial conditions:

$$u(x,0) = f(x)$$

$$u_t(x,0) = g(x)$$

The result is d'Alembert's formula:

$$u(x,t) = \frac{f(x-ct) + f(x+ct)}{2} + \frac{1}{2c}\int_{x-ct}^{x+ct} g(s)ds$$

In the classical sense if $f(x) \in C^k$ and $g(x) \in C^{k-1}$ then $u(t,x) \in C^k$. However, the waveforms F and G may also be generalized functions, such as the delta-function. In that case, the solution may be interpreted as an impulse that travels to the right or the left.

The basic wave equation is a linear differential equation and so it will adhere to the superposition principle. This means that the net displacement caused by two or more waves is the sum of the displacements which would have been caused by each wave individually. In addition, the behaviour of a wave can be analysed by breaking up the wave into components, e.g. the Fourier transform breaks up a wave into sinusoidal components.

Scalar Wave Equation in Three Space Dimensions

The solution of the initial-value problem for the wave equation in three space dimensions can be obtained from the solution for a spherical wave. This result can then be used to obtain the solution in two space dimensions.

Spherical Waves

The wave equation is unchanged under rotations of the spatial coordinates, because the Laplacian operator is invariant under rotation, and therefore one may expect to find solutions that depend only on the radial distance from a given point. Such solutions must satisfy

$$u_{tt} - c^2\left(u_{rr} + \frac{2}{r}u_r\right) = 0.$$

This equation may be rewritten as

$$(ru)_{tt} - c^2(ru)_{rr} = 0;$$

the quantity ru satisfies the one-dimensional wave equation. Therefore there are solutions in the form

$$u(t,r) = \frac{1}{r}F(r-ct) + \frac{1}{r}G(r+ct),$$

where F and G are arbitrary functions. Each term may be interpreted as a spherical wave that expands or contracts with velocity c. Such waves are generated by a point source, and they make possible sharp signals whose form is altered only by a decrease in amplitude as r increases. Such waves exist only in cases of space with odd dimensions.

Monochromatic Spherical Wave

A point source is vibrating at a single frequency f with phase = 0 at $t = 0$ with a peak-to-peak magnitude of $2a$. A spherical wave is propagated from the point. The phase of the propagated wave changes as kr where r is the distance travelled from the source. The magnitude falls off as $1/r$ since the energy falls off as $1/r^2$. The amplitude of the spherical wave at is therefore given by:

$$u(t,r) = Re\left[\frac{a}{r}e^{i(\omega t - kr)}\right]$$

Solution of a General Initial-Value Problem

The wave equation is linear in u and it is left unaltered by translations in space and time. Therefore we can generate a great variety of solutions by translating and summing spherical waves. Let

$\varphi(\partial,\eta,\zeta)$ be an arbitrary function of three independent variables, and let the spherical wave form F be a delta-function: that is, let F be a weak limit of continuous functions whose integral is unity, but whose support (the region where the function is non-zero) shrinks to the origin. Let a family of spherical waves have centre at (∂,η,ζ), and let r be the radial distance from that point. Thus

$$r^2 = (x-\xi)^2 + (y-\eta)^2 + (z-\zeta)^2.$$

If u is a superposition of such waves with weighting function φ, then

$$u(t,x,y,z) = \frac{1}{4\pi c}\iiint \varphi(\xi,\eta,\zeta)\frac{\delta(r-ct)}{r}\,d\xi\, d\eta\, d\zeta;$$

the denominator $4\pi c$ is a convenience.

From the definition of the delta-function, u may also be written as

$$u(t,x,y,z) = \frac{t}{4\pi}\iint_S \varphi(x+ct\alpha, y+ct\beta, z+ct\gamma)d\omega,$$

where α, β, and γ are coordinates on the unit sphere S, and ω is the area element on S. This result has the interpretation that $u(t,x)$ is t times the mean value of φ on a sphere of radius ct centred at x:

$$u(t,x,y,z) = tM_{ct}[\phi].$$

It follows that

$$u(0,x,y,z) = 0, \quad u_t(0,x,y,z) = \phi(x,y,z).$$

The mean value is an even function of t, and hence if

$$v(t,x,y,z) = \frac{\partial}{\partial t}\left(tM_{ct}[\psi]\right),$$

then

$$v(0,x,y,z) = \psi(x,y,z), \quad v_t(0,x,y,z) = 0.$$

These formulas provide the solution for the initial-value problem for the wave equation. They show that the solution at a given point P, given (t,x,y,z) depends only on the data on the sphere of radius ct that is intersected by the light cone drawn backwards from P. It does *not* depend upon data on the interior of this sphere.

Thus the interior of the sphere is a lacuna for the solution. This phenomenon is called Huygens' principle. It is true for odd numbers of space dimension, where for one dimension the integration is performed over the boundary of an interval with respect to the Dirac measure. It is not satisfied in even space dimensions. The phenomenon of lacunas has been extensively investigated in Atiyah, Bott and Gårding (1970, 1973).

Scalar Wave Equation in Two Space Dimensions

In two space dimensions, the wave equation is

$$u_{tt} = c^2\left(u_{xx} + u_{yy}\right).$$

We can use the three-dimensional theory to solve this problem if we regard u as a function in three dimensions that is independent of the third dimension. If

$$u(0,x,y) = 0, \quad u_t(0,x,y) = \phi(x,y),$$

then the three-dimensional solution formula becomes

$$u(t,x,y) = tM_{ct}[\phi] = \frac{t}{4\pi}\iint_S \phi(x + ct\alpha, y + ct\beta)d\omega,$$

where α and β are the first two coordinates on the unit sphere, and dω is the area element on the sphere. This integral may be rewritten as an integral over the disc D with centre (x,y) and radius ct:

$$u(t,x,y) = \frac{1}{2\pi c}\iint_D \frac{\phi(x+\xi, y+\eta)}{\sqrt{(ct)^2 - \xi^2 - \eta^2}}d\xi d\eta.$$

It is apparent that the solution at (t,x,y) depends not only on the data on the light cone where

$$(x-\xi)^2 + (y-\eta)^2 = c^2t^2,$$

but also on data that are interior to that cone.

Several Space Dimensions

The one-dimensional initial-boundary value theory may be extended to an arbitrary number of space dimensions. Consider a domain D in m-dimensional x space, with boundary B. Then the wave equation is to be satisfied if x is in D and $t > 0$. On the boundary of D, the solution u shall satisfy

$$\frac{\partial u}{\partial n} + au = 0,$$

where n is the unit outward normal to B, and a is a non-negative function defined on B. The case where u vanishes on B is a limiting case for a approaching infinity. The initial conditions are

$$u(0,x) = f(x), \quad u_t(0,x) = g(x),$$

where f and g are defined in D. This problem may be solved by expanding f and g in the eigenfunctions of the Laplacian in D, which satisfy the boundary conditions. Thus the eigenfunction v satisfies

$$\nabla \cdot \nabla v + \lambda v = 0,$$

in D, and

$$\frac{\partial v}{\partial n} + av = 0,$$

on B.

In the case of two space dimensions, the eigenfunctions may be interpreted as the modes of vibration of a drumhead stretched over the boundary B. If B is a circle, then these eigenfunctions have an angular component that is a trigonometric function of the polar angle θ, multiplied by a Bessel function (of integer order) of the radial component. Further details are in Helmholtz equation.

If the boundary is a sphere in three space dimensions, the angular components of the eigenfunctions are spherical harmonics, and the radial components are Bessel functions of half-integer order.

Inhomogeneous Wave Equation in One Dimension

The inhomogeneous wave equation in one dimension is the following:

$$c^2 u_{xx}(x,t) - u_{tt}(x,t) = s(x,t)$$

with initial conditions given by

$$u(x,0) = f(x)$$

$$u_t(x,0) = g(x)$$

The function $s(x,t)$ is often called the source function because in practice it describes the effects of the sources of waves on the medium carrying them. Physical examples of source functions include the force driving a wave on a string, or the charge or current density in the Lorenz gauge of electromagnetism.

One method to solve the initial value problem (with the initial values as posed above) is to take advantage of the property of the wave equation that its solutions obey causality. That is, for any point (x_i, t_i), the value of $u(x_i, t_i)$ depends only on the values of $f(x_i + ct_i)$ and $f(x_i - ct_i)$ and the values of the function $g(x)$ between $(x_i - ct_i)$ and $(x_i + ct_i)$. This can be seen in d'Alembert's formula, stated above, where these quantities are the only ones that show up in it. Physically, if the maximum propagation speed is c, then no part of the wave that can't propagate to a given point by a given time can affect the amplitude at the same point and time.

In terms of finding a solution, this causality property means that for any given point on the line being considered, the only area that needs to be considered is the area encompassing all the points that could causally affect the point being considered. Denote the area that casually affects point (x_i, t_i) as Rc. Suppose we integrate the inhomogeneous wave equation over this region.

$$\iint_{R_C} \left(c^2 u_{xx}(x,t) - u_{tt}(x,t)\right) dx dt = \iint_{R_C} s(x,t) dx dt.$$

To simplify this greatly, we can use Green's theorem to simplify the left side to get the following:

$$\int_{L_0+L_1+L_2} \left(-c^2 u_x(x,t)dt - u_t(x,t)dx\right) = \iint_{R_C} s(x,t) dx dt.$$

The left side is now the sum of three line integrals along the bounds of the causality region. These turn out to be fairly easy to compute

$$\int_{x_i-ct_i}^{x_i+ct_i} -u_t(x,0)dx = -\int_{x_i-ct_i}^{x_i+ct_i} g(x)dx.$$

In the above, the term to be integrated with respect to time disappears because the time interval involved is zero, thus $dt = 0$. For the other two sides of the region, it is worth noting that $x \pm ct$ is a constant, namingly $x_i \pm ct_i$, where the sign is chosen appropriately. Using this, we can get the relation $dx \pm cdt = 0$, again choosing the right sign:

$$\begin{aligned} &\int_{L_1} \left(-c^2 u_x(x,t)dt - u_t(x,t)dx\right) \\ &= \int_{L_1} \left(cu_x(x,t)dx + cu_t(x,t)dt\right) \\ &= c\int_{L_1} du(x,t) = cu(x_i,t_i) - cf(x_i + ct_i). \end{aligned}$$

And similarly for the final boundary segment:

$$\begin{aligned} &\int_{L_2} \left(-c^2 u_x(x,t)dt - u_t(x,t)dx\right) \\ &= -\int_{L_2} \left(cu_x(x,t)dx + cu_t(x,t)dt\right) \\ &= -c\int_{L_2} du(x,t) = -\left(cf(x_i - ct_i) - cu(x_i,t_i)\right) \\ &\qquad = cu(x_i,t_i) - cf(x_i - ct_i). \end{aligned}$$

Adding the three results together and putting them back in the original integral:

$$\begin{aligned} &-\int_{x_i-ct_i}^{x_i+ct_i} g(x)dx + cu(x_i,t_i) - cf(x_i + ct_i) + cu(x_i,t_i) - cf(x_i - ct_i) \\ &= \iint_{R_C} s(x,t) dx dt \end{aligned}$$

$$2cu(x_i,t_i)-\int_{x_i-ct_i}^{x_i+ct_i} g(x)dx-cf(x_i+ct_i)$$
$$-cf(x_i-ct_i)=\iint_{R_C} s(x,t)dxdt$$

$$2cu(x_i,t_i)=\int_{x_i-ct_i}^{x_i+ct_i} g(x)dx+cf(x_i+ct_i)$$
$$+cf(x_i-ct_i)+\iint_{R_C} s(x,t)dxdt$$

$$u(x_i,t_i)=\frac{f(x_i+ct_i)+f(x_i-ct_i)}{2}$$
$$+\frac{1}{2c}\int_{x_i-ct_i}^{x_i+ct_i} g(x)dx+\frac{1}{2c}\int_0^{t_i}\int_{x_i-c(t_i-t)}^{x_i+c(t_i-t)} s(x,t)dxdt.$$

In the last equation of the sequence, the bounds of the integral over the source function have been made explicit. Looking at this solution, which is valid for all choices (x_i,t_i) compatible with the wave equation, it is clear that the first two terms are simply d'Alembert's formula, as stated above as the solution of the homogeneous wave equation in one dimension. The difference is in the third term, the integral over the source.

Other Coordinate Systems

In three dimensions, the wave equation, when written in elliptic cylindrical coordinates, may be solved by separation of variables, leading to the Mathieu differential equation.

Wave Packet

In physics, a wave packet (or wave train) is a short "burst" or "envelope" of localized wave action that travels as a unit. A wave packet can be analysed into, or can be synthesized from, an infinite set of component sinusoidal waves of different wavenumbers, with phases and amplitudes such that they interfere constructively only over a small region of space, and destructively elsewhere. Depending on the evolution equation, the wave packet's envelope may remain constant or it may change (dispersion) while propagating. Quantum mechanics ascribes a special significance to the wave packet: it is interpreted as a "probability wave", describing the probability that a particle or particles in a particular state will be measured to have a given position and momentum.

It is in this way related to the wave function. Through application of the Schrödinger equation in quantum mechanics, it is possible to deduce the time evolution of a system, similar to the process of the Hamiltonian formalism in classical mechanics. The wave packet is thus a mathematical solution to the Schrödinger equation. The area under the absolute square of the wave packet solution is interpreted as the probability density of finding the particle in a region. The dispersive character of solutions of the Schrödinger equation has played an important role in rejecting Schrödinger's original interpretation, and accepting the Born rule.

In the coordinate representation of the wave (such as the Cartesian coordinate system), the localized position of the physical object's probability is given by the position of the packet. Moreover, the narrower the spatial wave packet, and therefore the better defined the position of the wave packet, the larger the spread in the momentum of the wave. This trade-off between spread in position and spread in momentum is one example of the Heisenberg uncertainty principle.

Background

In the early 1900s, it became apparent that classical mechanics had some major failings. Isaac Newton originally proposed the idea that light came in discrete packets, which he called "corpuscles", but the wave-like behaviour of many light phenomena quickly led scientists to favour a wave description of electromagnetism. It wasn't until the 1930s that the particle nature of light really began to be widely accepted in physics. The development of quantum mechanics — and its success at explaining confusing experimental results — was at the root of this acceptance. Thus, one of the basic concepts in the formulation of quantum mechanics is that of light coming in discrete bundles called photons. The energy of light photon is a discrete function of its frequency,

$$E = nh\nu \ .$$

The energy is a positive integer, n, multiple of Planck's constant, h, and frequency, ν. (This resolved a significant problem in classical physics, called the ultraviolet catastrophe.)

The ideas of quantum mechanics continued to be developed throughout the 20th century. The picture that was developed was of a particulate world, with all phenomena and matter made of and interacting with discrete particles; however, these particles were described by a probability wave. The interactions, locations, and all of physics would be reduced to the calculations of these probability amplitude waves. The particle-like nature of the world was significantly confirmed by experiment, while, at the very same time, the wave-like phenomena could be characterised as consequences of the wave-packet nature of particles.

Basic Behaviours of Wave Packets

As an example of propagation *without dispersion*, consider wave solutions to the following wave equation:

$$\frac{\partial^2 u}{\partial t^2} = c^2 \nabla^2 u \,,$$

where c is the speed of the wave's propagation in a given medium. Using the physics time convention, exp("$i\omega t$), the wave equation has plane-wave solutions

$$u(\mathbf{x},t) = e^{i(\mathbf{k}\cdot\mathbf{x}-\omega t)} \,,$$

where

$$\omega^2 = |\,\mathbf{k}\,|^2\, c^2, |\,\mathbf{k}\,|^2 = k_x^2 + k_y^2 + k_z^2 .$$

This relation between ω and k should be valid so that the plane wave is a solution to the wave equation. It is called a dispersion relation.

To simplify, consider only waves propagating in one dimension (extension to three dimensions is straightforward). Then the general solution is

$$u(x,t) = Ae^{i(kx-\omega t)} + Be^{-i(kx+\omega t)} \,,$$

in which we may take $\omega = kc$. The first term represents a wave propagating in the positive x-direction since it is a function of $x - ct$ only; the second term, being a function of $x + ct$, represents a wave propagating in the negative x-direction.

A wave packet is a localized disturbance that results from the sum of many different wave forms. If the packet is strongly localized, more frequencies are needed to allow the constructive superposition in the region of localization and destructive superposition outside the region. From the basic solutions in one dimension, a general form of a wave packet can be expressed as

$$u(x,t) = \frac{1}{\sqrt{2\pi}} \int_{-\infty}^{\infty} A(k) e^{i(kx-\omega(k)t)}\, dk \,.$$

As in the plane-wave case the wave packet travels to the right for $\omega(k) = kc$ (since then $u(x,t) = F(x-ct)$) and to the left for $\omega(k) = -kc$ (since then $u(x,t) = F(x+ct)$).

The factor $1/\sqrt{2\pi}$ comes from Fourier transform conventions. The amplitude *A(k)* contains the coefficients of the linear superposition of the plane-wave solutions. These coefficients can in turn be expressed as a function of *u(x,t)* evaluated at $t = 0$ by inverting the Fourier transform relation above:

$$A(k) = \frac{1}{\sqrt{2\pi}} \int_{-\infty}^{\infty} u(x,0) e^{-ikx}\, dx .$$

For instance, choosing

$$u(x,0) = e^{-x^2 + ik_0 x},$$

we obtain

$$A(k) = \frac{1}{\sqrt{2}} e^{-\frac{(k-k_0)^2}{4}},$$

and finally

$$u(x,t) = e^{-(x-ct)^2 + ik_0(x-ct)} .$$

The nondispersive propagation of the real or imaginary part of this wave packet is presented in the above animation.

As an example of propagation *with dispersion*, consider solutions to the Schrödinger equation (with *m* and ‘ set equal to one)

$$i\frac{\partial u}{\partial t} = -\frac{1}{2}\nabla^2 u,$$

yielding as dispersion relation

$$\omega = \frac{1}{2} |\mathbf{k}|^2 .$$

Once again, restricting ourselves to one dimension the solution to the Schrödinger equation satisfying the initial condition $u(x,0) = e^{-x^2 + ik_0 x}$ is found according to

$$u(x,t) = \frac{e^{-\frac{k_0^2}{4}}}{\sqrt{1+2it}} e^{-\frac{(x-\frac{ik_0}{2})^2}{1+2it}} .$$

An impression of the dispersive behaviour of this wave packet is obtained by looking at

$$|u(x,t)| = \frac{1}{(1+4t^2)^{1/4}} e^{-\frac{(x-k_0 t)^2}{1+4t^2}}$$

(note that $|u(x,t)|$ itself is not a solution of the Schrödinger equation).

It is evident that this dispersive wave packet, while moving with constant group velocity k_0, has a width increasing with time as $(1+4t^2)^{1/2}$.

Gaussian Wavepackets in Quantum Mechanics

The above Gaussian wavepacket, unnormalized and just centred at the origin, instead, can now be written in 3D:

$$\psi(x) = e^{-\mathbf{r}\cdot\mathbf{r}/2a},$$

where a is a positive real number, the *square of the width of the wavepacket*, $a = 2\langle \mathbf{r}\cdot\mathbf{r}\rangle / 3\langle 1\rangle = 2 \ (\Delta x)^2$.

The Fourier transform is also a Gaussian in terms of the wavenumber, the k-vector, (with inverse width, $1/a = 2\langle \mathbf{k}\cdot\mathbf{k}\rangle / 3\langle 1\rangle = 2\left(\Delta p_x / \hbar\right)^2$, so that $\Delta x \, \Delta p_x = \hbar / 2$, i.e. it saturates the uncertainty relation),

$$\psi(\mathbf{k}) = (2\pi a)^{3/2} e^{-a\mathbf{k}\cdot\mathbf{k}/2}.$$

Each separate wave only phase-rotates in time, so that the time dependent Fourier-transformed solution is:

$$\begin{aligned}\Psi(\mathbf{k},t) &= (2\pi a)^{3/2} e^{-a\mathbf{k}\cdot\mathbf{k}/2} e^{-iEt/\hbar} \\ &= (2\pi a)^{3/2} e^{-a\mathbf{k}\cdot\mathbf{k}/2 - i(\hbar^2\mathbf{k}\cdot\mathbf{k}/2m)t/\hbar} \\ &= (2\pi a)^{3/2} e^{-(a+i\hbar t/m)\mathbf{k}\cdot\mathbf{k}/2}.\end{aligned}$$

The inverse Fourier transform is still a Gaussian, but the parameter a has become complex, and there is an overall normalization factor.

$$\Psi(\mathbf{r},t) = \left(\frac{a}{a + i\hbar t / m}\right)^{3/2} e^{-\frac{\mathbf{r}\cdot\mathbf{r}}{2(a+i\hbar t/m)}}.$$

The integral of Ø over all space is invariant, because it is the inner product of Ø with the state of zero energy, which is a wave with infinite wavelength, a constant function of space. For any energy eigenstate $\eta(x)$, the inner product:

$$\langle \eta | \psi \rangle = \int \eta(\mathbf{r})\psi(\mathbf{r}) d^3\mathbf{r},$$

only changes in time in a simple way: its phase rotates with a frequency determined by the energy of η. When η has zero energy, like the infinite wavelength wave, it doesn't change at all. The quantity $|Ø|^2$ is also invariant, which is a statement of the conservation of probability. Explicitly,

$$|\Psi|^2 - \Psi^*\Psi = \frac{a}{\sqrt{a^2 + (\hbar t / m)^2}} e^{-\frac{\mathbf{r}\cdot\mathbf{r}a}{a^2+(\hbar t/m)^2}}.$$

The width of the Gaussian is the interesting quantity which can be read off from $|Ø|^2$:

$$\sqrt{\frac{a^2+(\hbar t/m)^2}{a}}.$$

The width eventually grows linearly in time, as ‘t /m– a. This is wave-packet spreading: no matter how narrow the initial wavefunction, a Schrödinger wave eventually fills all of space.

This linear growth is a reflection of the momentum uncertainty — the wavepacket is confined to a narrow width –a, and so has a momentum which is uncertain (according to the uncertainty principle) by the amount ‘/2–a, a spread in velocity of ‘/2m–a, and thus in the future position by ‘t /m–a. (The uncertainty relation is then a strict inequality, far from saturation.)

Free Propagator

The narrow-width limit of the Gaussian wavepacket solution discussed is the free *propagator kernel* *K*. For other differential equations, this is sometimes called the Green's function, but in quantum mechanics it is traditional to reserve the name Green's function for the time Fourier transform of *K*.

Returning to one dimension for simplicity, when *a* is the infinitesimal quantity ε, the Gaussian initial condition, rescaled so that its integral is one:

$$\psi_0(x)=\frac{1}{\sqrt{2\pi\epsilon}}e^{-\frac{x^2}{2\epsilon}}$$

becomes a delta function, δ(*x*), so that its time evolution,

$$K_t(x)=\frac{1}{\sqrt{2\pi(it+\epsilon)}}e^{\frac{-x^2}{2it+\epsilon}}$$

gives the *propagator*.

Note that a very narrow initial wavepacket instantly becomes infinitely wide, but with a phase which is more rapidly oscillatory at large values of *x*. This might seem strange—the solution goes from being localized at one point to being "everywhere" at *all later times*, but it is a reflection of the enormous momentum uncertainty of a localized particle. Also note that the norm of the wavefunction is infinite, but this is also correct, since the square of a delta function is divergent in the same way.

The factor involving ε is an infinitesimal quantity which is there to make sure that integrals over *K* are well defined. In the limit that ε→0, *K* becomes purely oscillatory, and integrals of *K* are not absolutely convergent. In the remainder of this section, it *will* be set to zero, but in order for all the integrations over intermediate states to be well defined, the limit $\varepsilon \to 0$ is to be only taken after the final state is calculated. The propagator is the amplitude for reaching point x at time t, when starting at the origin, $x = 0$. By translation invariance, the amplitude for reaching a point x when starting at point y is the same function, only now translated,

$$K_t(x,y) = K_t(x-y) = \frac{1}{\sqrt{2\pi i t}} e^{\frac{-i(x-y)^2}{2t}} .$$

In the limit when t is small, the propagator, of course, goes to a delta function,

$$\lim_{t\to 0} K_t(x-y) = \delta(x-y) \, ,$$

but only in the sense of distributions: The integral of this quantity multiplied by an arbitrary differentiable test function gives the value of the test function at zero. To see this, note that the integral over all space of K equals 1 at all times:

$$\int_x K_t(x) = 1,$$

since this integral is the inner-product of K with the uniform wavefunction. But the phase factor in the exponent has a nonzero spatial derivative everywhere except at the origin, and so when the time is small there are fast phase cancellations at all but one point. This is rigorously true when the limit $\varepsilon \to 0$ is taken at the very end.

So the propagation kernel is the (future) time evolution of a delta function, and it is continuous, in a sense: it goes to the initial delta function at small times. If the initial wavefunction is an infinitely narrow spike at position y,

$$\psi_0(x) = \delta(x-y),$$

it becomes the oscillatory wave:

$$\psi_t(x) = \frac{1}{\sqrt{2\pi i t}} e^{-i(x-y)^2/2t} .$$

Now, since every function can be written as a weighted sum of such narrow spikes,

$$\psi_0(x) = \int_y \psi_0(y)\delta(x-y),$$

the time evolution of *every function* ψ_0 is determined by the propagation kernel K:

$$\psi_t(x) = \int_y \psi_0(x) \frac{1}{\sqrt{2\pi i t}} e^{-i(x-y)^2/2t}.$$

Thus, this is a formal way to express the general solution. The interpretation of this expression is that the amplitude for a particle to be found at point x at time t is the amplitude that it started at y, times the amplitude that it went from y to x, *summed over all the possible starting points*. In other words, it is a convolution of the kernel K with the arbitrary initial condition ψ_0,

$$\psi_t = K * \psi_0.$$

Since the amplitude to travel from x to y after a time $t+t'$ can be considered in two steps, the propagator obeys the identity:

$$\int_y K(x-y;t)K(y-z;t') = K(x-z;t+t'),$$

which can be interpreted as follows: the amplitude to travel from x to z in time $t+t'$ is the sum of the amplitude to travel from x to y in time t, multiplied by the amplitude to travel from y to z in time t', summed over *all possible intermediate states y*. This is a property of an arbitrary quantum system, and by subdividing the time into many segments, it allows the time evolution to be expressed as a path integral.

Analytic Continuation to Diffusion

The spreading of wavepackets in quantum mechanics is directly related to the spreading of probability densities in diffusion. For a particle which is randomly walking, the probability density function at any point satisfies the diffusion equation:

$$\frac{\partial}{\partial t}\rho = \frac{1}{2}\frac{\partial^2}{\partial x^2}\rho,$$

where the factor of 2, which can be removed by a rescaling either time or space, is only for convenience.

A solution of this equation is the spreading Gaussian,

$$\rho_t(x) = \frac{1}{\sqrt{2\pi t}} e^{\frac{-x^2}{2t}},$$

and since the integral of ρ_t *is constant, while the width is becoming narrow at small times, this function approaches a delta function at* t = *0:*

$$\lim_{t\to 0}\rho_t(x) = \delta(x)\mathbb{S}$$

again only in the sense of distributions, so that

$$\lim_{t\to 0}\int_x f(x)\rho_t(x) = f(0)$$

for any smooth test function f.

The spreading Gaussian is the propagation kernel for the diffusion equation and it obeys the convolution identity:

$$K_{t+t'} = K_t * K_{t'},$$

which allows diffusion to be expressed as a path integral. The propagator is the exponential of an operator H:

$$K_t(x) = e^{-tH},$$

which is the infinitesimal diffusion operator,

$$H = -\frac{\nabla^2}{2}.$$

A matrix has two indices, which in continuous space makes it a function of x and x ‘. In this case, because of translation invariance, the matrix element K only depend on the difference of the position, and a convenient abuse of notation is to refer to the operator, the matrix elements, and the function of the difference by the same name:

$$K_t(x,x') = K_t(x-x').$$

Translation invariance means that continuous matrix multiplication:

$$C(x,x'') = \int_{x'} A(x,x')B(x',x''),$$

is essentially convolution,

$$C(\Delta) = C(x-x'') = \int_{x'} A(x-x')B(x'-x'') = \int_y A(\Delta-y)B(y).$$

The exponential can be defined over a range of ts which include complex values, so long as integrals over the propagation kernel stay convergent,

$$K_z(x) = e^{-zH}.$$

As long as the real part of z is positive, for large values of x, K is exponentially decreasing, and integrals over K are indeed absolutely convergent.

The limit of this expression for z coming close to the pure imaginary axis is the Schrödinger propagator:

$$K_t^{\text{Schr}} = K_{it+\epsilon} = e^{-(it+\epsilon)H},$$

and this gives a more conceptual explanation for the time evolution of Gaussians.

From the fundamental identity of exponentiation, or path integration:

$$K_z * K_{z'} = K_{z+z'}$$

holds for all complex z values where the integrals are absolutely convergent so that the operators are well defined.

Thus, quantum evolution starting from a Gaussian, which is the diffusion kernel K,

$$\psi_0(x) = K_a(x) = K_a * \delta(x)$$

gives the time evolved state,

$$\psi_t = K_{it} * K_a = K_{a+it}.$$

This illustrates the above diffusive form of the Gaussian solutions,

$$\psi_t(x) = \frac{1}{\sqrt{2\pi(a+it)}} e^{-\frac{x^2}{2(a+it)}}.$$

Chapter 2

Virtual Particle

In physics, a virtual particle is a particle that exists for a limited time and space. The energy and momentum of a virtual particle are uncertain according to the uncertainty principle. The degree of uncertainty of each is inversely proportional to time duration (for energy) or to position span (formomentum).

Virtual particles exhibit some of the phenomena that real particles do, such as obedience to the conservation laws. If a single particle is detected, then the consequences of its existence are prolonged to such a degree that it cannot be virtual. Virtual particles are viewed as the quanta that describe fields of the basic force interactions, which cannot be described in terms of real particles. Examples of these are static force fields, such as a simple electric or magnetic field, or the components of any field that do not carry information from place to place at the speed of light (information radiated by means of a field must be composed of real particles). Virtual photons are also a major component of antenna near field phenomena and induction fields, which have shorter-range effects, and do not radiate through space with the same range-properties as do electromagnetic wavephotons. For example, the energy carried from one winding of a transformer to another, or to and from a patient in an MRI scanner, in quantum terms is carried by virtual photons, not real photons.

The virtual particle forms of massless particles, such as photons, do have mass (which may be either positive or negative) and are said to be off mass shell. They are allowed to have mass (which consists of "borrowed energy") because they exist for only a temporary time, which in turn gives them a limited "range". This is in accordance with the uncertainty principle which allows existence of such particles of borrowed energy, so long as their energy, multiplied by the time they exist, is a fraction of Planck's constant. Possession of mass also allows single virtual photons to be more easily created and emitted from single charged elementary particles, something that cannot happen for massless photons without violating conservation of momentum and energy (single real photons are always created and emitted from systems of two or more particles). For particles which do have a rest mass, their virtual forms still violate the energy-momentum relation of special relativity, in having a mass more or less than predicted by the relation:

$$E^2 - p^2 = m^2$$

The concept of virtual particles is closely related to the idea of quantum fluctuations. Virtual particles can be thought of as coming into existence as quantities, such as the electric field, which fluctuate around their expectation values as required by quantum mechanics.

Properties

The concept of virtual particles arises in the perturbation theory of quantum field theory, an approximation scheme in which interactions (essentially forces) between real particles are calculated in terms of exchanges of virtual particles. Any process involving virtual particles admits a schematic representation known as a Feynman diagram which facilitates the understanding of calculations.

A virtual particle is one that does not precisely obey the $m^2c^4 = E^2$ " p^2c^2 relationship for a short time. In other words, its kinetic energy may not have the usual relationship to velocity–indeed, it can be negative. The probability amplitude for it to exist tends to be canceled out by destructive interference over longer distances and times. A virtual particle can be considered a manifestation ofquantum tunnelling. The range of forces carried by virtual particles is limited by the uncertainty principle, which regards energy and time as conjugate variables; thus virtual particles of larger mass have more limited range.

There is not a definite line differentiating virtual particles from real particles — the equations of physics just describe particles (which includes both equally). The amplitude that a virtual particle exists interferes with the amplitude for its non-existence; whereas for a real particle the cases of existence and non-existence cease to be coherent with each other and do not interfere any more. In the quantum field theory view, "real particles" are viewed as being detectable excitations of underlying quantum fields. As such, virtual particles are also excitations of the underlying fields, but are detectable only as forces but not particles. They are "temporary" in the sense that they appear in calculations, but are not detected as single particles. Thus, in mathematical terms, they never appear as indices to the scattering matrix, which is to say, they never appear as the observable inputs and outputs of the physical process being modelled. In this sense, virtual particles are an artifact of perturbation theory, and do not appear in a non-perturbative treatment.

There are two principal ways in which the notion of virtual particles appears in modern physics. They appear as intermediate terms in Feynman diagrams; that is, as terms in a perturbative calculation. They also appear as an infinite set of states to be summed or integrated over in the calculation of a semi-non-perturbative effect. In the latter case, it is sometimes said that virtual particles cause the effect, or that the effect occurs because of the existence of virtual particles.

Manifestations

There are many observable physical phenomena resulting from interactions involving virtual particles. For bosonic particles which exhibit rest mass when they are free and "real," virtual interactions are characterized by the relatively short range of the force interaction produced by particle exchange. Examples of such short-range interactions are the strong and weak forces, and their associated field bosons. For the gravitational and electromagnetic forces, the zero rest-mass of the associated boson particle permits long-range forces to be mediated by virtual particles. However, in the case of photons, power and information transfer by virtual particles is a relatively short-range phenomenon (existing importantly only within a few wavelengths of the field-disturbance which carries information or transferred power), as for example seen in the characteristically short range of inductive and capacitative effects in the near field zone of coils and antennas.

Some field interactions which may be seen in terms of virtual particles are:

- The Coulomb force (static electric force) between electric charges. It is caused by the exchange of virtual photons. In symmetric 3-dimensional space this exchange results in the inverse square law for electric force. Since the photon has no mass, the coulomb potential has an infinite range.

- The magnetic field between magnetic dipoles. It is caused by the exchange of virtual photons. In symmetric 3-dimensional space this exchange results in the inverse square law for magnetic force. Since the photon has no mass, the magnetic potential has an infinite range.
- Much of the so-called near field of radio antennas, where the magnetic and electric effects of the changing current in the antenna wire and the charge effects of the wire's capacitive charge are important contributors to the total EM field, but both of which effects decay with increasing distance from the antenna much more quickly than do the influence of "conventional" electromagnetic waves far from the source (in terms of wavelength), for which E is always equal to cB, and which are composed of real photons.

  It should be noted that real and virtual photons are mixed near an antenna, with the virtual photons responsible only for the "extra" magnetic-inductive and static-electric effects which cause an imbalance between E and cB. As distance from the antenna grows, the near-field effects die out more quickly, and only the "radiative" effects due to real photons remain as important effects.

  Virtual effects extend to infinity, but drop off in field strength as $1/r^2$ rather than the field of EM waves composed of real photons, which drop $1/r$ (the powers, respectively, decrease as $1/r^4$ and $1/r^2$). See near and far field for a more detailed discussion. See near field communication for practical communications applications.
- Electromagnetic induction. This phenomenon transferring energy to and from a magnetic coil via a changing (electro)magnetic field can be viewed as a near-field effect. It is the basis for power transfer in transformers and electric generators and motors, and also signal transfer in metal detectors, magnetic and magnetoacustic anti theft electronic tags, and even signals between patient and machine in an MRI scanner.

  Some confusion about the use of "radio waves" results when these devices are used at conventional RF frequencies, as they are in an MRI scanner.See resonant inductive coupling and wireless energy transfer for other practical examples.

- The strong nuclear force between quarks is the result of interaction of virtual gluons. The residual of this force outside of quark triplets (neutron and proton) holds neutrons and protons together in nuclei, and is due to virtual mesons such as the pi meson and rho meson.
- The weak nuclear force - it is the result of exchange by virtual W and Z bosons.
- The spontaneous emission of a photon during the decay of an excited atom or excited nucleus; such a decay is prohibited by ordinary quantum mechanics and requires the quantization of the electromagnetic field for its explanation.
- The Casimir effect, where the ground state of the quantized electromagnetic field causes attraction between a pair of electrically neutral metal plates.
- The van der Waals force, which is partly due to the Casimir effect between two atoms,
- Vacuum polarization, which involves pair production or the decay of the vacuum, which is the spontaneous production of particle-antiparticle pairs (such as electron-positron).
- Lamb shift of positions of atomic levels.
- Hawking radiation, where the gravitational field is so strong that it causes the spontaneous production of photon pairs (with black body energy distribution) and even of particle pairs.

Most of these have analogous effects in solid-state physics; indeed, one can often gain a better intuitive understanding by examining these cases. In semiconductors, the roles of electrons, positrons and photons in field theory are replaced by electrons in the conduction band, holes in the valence band, and phonons or vibrations of the crystal lattice. A virtual particle is in a virtual state where the probability amplitude is not conserved. Examples of macroscopic virtual phonons, photons, and electrons in the case of the tunnelling process were presented by Günter Nimtz in and.

History

Paul Dirac was the first to propose that empty space (a vacuum) can be visualized as consisting of a sea of virtual electron-positron pairs, known as the Dirac sea. The Dirac sea has a direct analog to the electronic band structure in crystalline solids as described in solid state physics. Here, particles correspond to conduction electrons, and antiparticles to holes. A variety of interesting phenomena can be attributed to this structure.

Virtual Particles in Feynman Diagrams

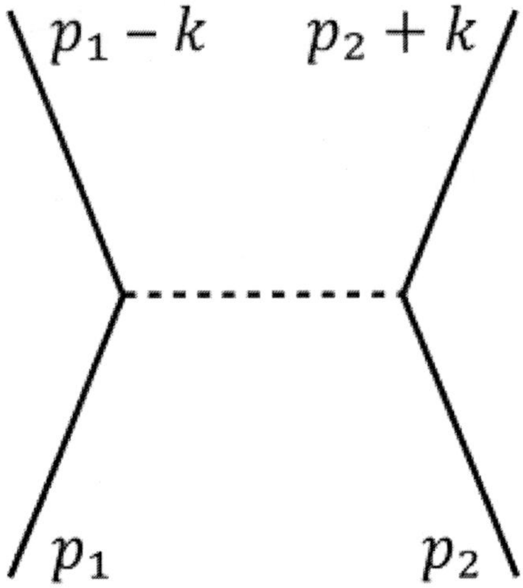

Figure: *One particle exchange scattering diagram*

The calculation of scattering amplitudes in theoretical particle physics requires the use of some rather large and complicated integrals over a large number of variables. These integrals do, however, have a regular structure, and may be represented as Feynman diagrams. The appeal of the Feynman diagrams is strong, as it allows for a simple visual presentation of what would otherwise be a rather arcane and abstract formula. In particular, part of the appeal is that the outgoing legs of a Feynman diagram can be associated with real, on-shell particles. Thus, it is natural to associate the other lines in the diagram with particles as well, called the “virtual particles”. Mathematically, they correspond to the propagators appearing in the diagram.

In the image to the right, the solid lines correspond to real particles (of momentum p_1 and so on), while the dotted line corresponds to a virtual particle carrying momentum k. For example, if the solid lines were to correspond to electrons interacting by means of the electromagnetic interaction, the dotted line would correspond to the exchange of a virtual photon. In the case of interacting nucleons, the dotted line would be a virtual pion. In the case of quarks interacting by means of the strong force, the dotted line would be a virtual gluon, and so on.

It is sometimes said that all photons are virtual photons. This is because the world-lines of photons always resemble the dotted line in the above Feynman diagram: the photon was emitted somewhere (say, a distant star), and then is absorbed somewhere else (say a photoreceptor cell in the eyeball). Furthermore, in a vacuum, a photon experiences no passage of (proper) time between emission and absorption. This statement illustrates the difficulty of trying to distinguish between "real" and "virtual" particles as mathematically they are the same objects and it is only our definition of "reality" which is weak here. In practice, a clear distinction can be made: real photons are detected as individual particles in particle detectors, whereas virtual photons are not directly detected; only their average or side-effects may be noticed, in the form of *forces* or (in modern language) interactions between particles.

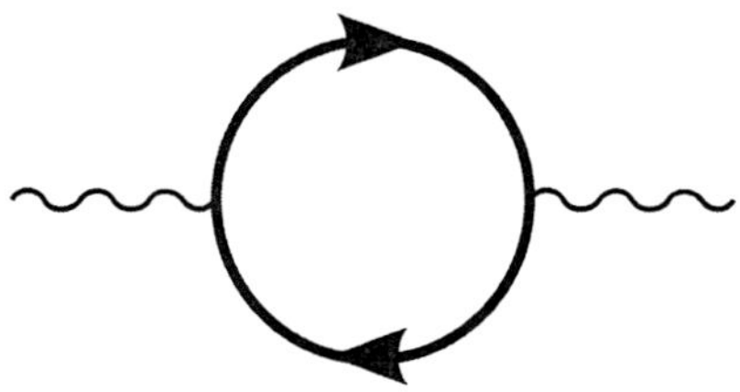

Figure: *One-loop diagram with fermion propagator*

Virtual particles need to be mesons or vector bosons, as in the example above; they may also befermions. However, in order to preserve quantum numbers, most simple diagrams involving fermion exchange are prohibited. The image to the right shows an allowed diagram, a one-loop diagram. The solid lines correspond to a fermion propagator, the wavy lines to bosons.

Virtual Particles in Vacuums

Formally, a particle is considered to be an eigenstate of the particle number operator $a^{\dagger}a$ where a is the particle annihilation operator and $a^{\dagger}$ the particle creation operator (sometimes collectively called ladder operators). In many cases, the particle number operator does not commute with the Hamiltonian for the system. This implies the number of particles in an area of space is not a well-defined quantity but like other quantum observables is represented by a probability distribution. Since these particles do not have a permanent existence, they are called *virtual particles* or vacuum fluctuations of vacuum energy. In a certain sense, they can be understood to be a manifestation of the time-energy uncertainty principle in a vacuum.

An important example of the "presence" of virtual particles in a vacuum is the Casimir effect. Here, the explanation of the effect requires that the total energy of all of the virtual particles in a vacuum can be added together. Thus, although the virtual particles themselves are not directly observable in the laboratory, they do leave an observable effect: their zero-point energy results in forces acting on suitably arranged metal plates or dielectrics.

Pair Production

In order to conserve the total fermion number of the universe, a fermion cannot be created without also creating its antiparticle; thus many physical processes lead to pair creation. The need for the normal ordering of particle fields in the vacuum can be interpreted by the idea that a pair of virtual particles may briefly "pop into existence", and then annihilate each other a short while later.

Thus, virtual particles are often popularly described as coming in pairs, a particle and antiparticle, which can be of any kind. These pairs exist for an extremely short time, and mutually annihilate in short order. In some cases, however, it is possible to boost the pair apart using external energy so that they avoid annihilation and become real particles. This may occur in one of two ways. In an accelerating frame of reference, the virtual particles may appear to be real to the accelerating observer; this is known as the Unruh effect. In short, the vacuum of a stationary frame appears, to the accelerated observer, to be a warm gas of real particles inthermodynamic equilibrium. The Unruh effect is a toy model for understanding Hawking radiation, the process by which black holes evaporate.

Another example is pair production in very strong electric fields, sometimes called vacuum decay. If, for example, a pair of atomic nuclei are merged together to very briefly form a nucleus with a charge greater than about 140, (that is, larger than about the inverse of the fine structure constant), the strength of the electric field will be such that it will be energetically favourable to create positron-electron pairs out of the vacuum or Dirac sea, with the electron attracted to the nucleus to annihilate the positive charge. This pair-creation amplitude was first calculated by Julian Schwinger in 1951.

The restriction to particle–antiparticle pairs is actually only necessary if the particles in question carry a conserved quantity, such as electric charge, which is not present in the initial or final state. Otherwise, other situations can arise. For instance, the beta decay of a neutron can happen through the emission of a single virtual, negatively charged W particle that almost immediately decays into a real electron and antineutrino; the neutron turns into a proton when it emits the W particle. The evaporation of a black hole is a process dominated by photons, which are their own antiparticles and are uncharged.

Cyclotron

In technology, a cyclotron is a type of particle accelerator. In physics, the cyclotron frequency or gyrofrequency is the frequency of a charged particle moving perpendicularly to the direction of a uniform magnetic field, i.e. a magnetic field of constant magnitude and direction.

Since that motion is always circular, the cyclotron frequency is well defined. Cyclotrons accelerate charged particles using a high-frequency, alternating voltage (potential difference). A perpendicular magnetic field causes the particles to spiral almost in a circle so that they re-encounter the accelerating voltage many times.

Figure: *A French cyclotron, produced in Zurich, Switzerland in 1937*

The cyclotron was invented by Leo Szilárd and first manufactured by Ernest Lawrence, of the University of California, Berkeley who started operating it in 1932. Others had been working along similar lines at the time. The first European cyclotron was constructed in Leningrad in the physics department of the Radium Institute, headed by Vitali Khlopin. This instrument was first proposed in 1932 by George Gamow and Lev Mysovskii and was installed and running by 1937.

TRIUMF, Canada's national laboratory for nuclear and particle physics, houses the world's largest cyclotron. The 18m diametre, 4,000 tonne main magnet produces a field of 0.46 T while a 23 MHz 94 kV electric field is used to accelerate the 300 μA beam. TRIUMF is run by a consortium of sixteen Canadian universities and is located at the University of British Columbia, Vancouver, Canada.

How the Cyclotron Works

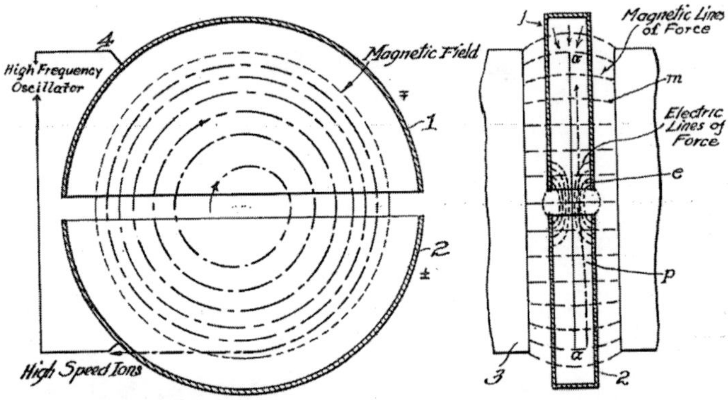

Figure: *Diagram of cyclotron operation from Lawrence's 1934 patent*

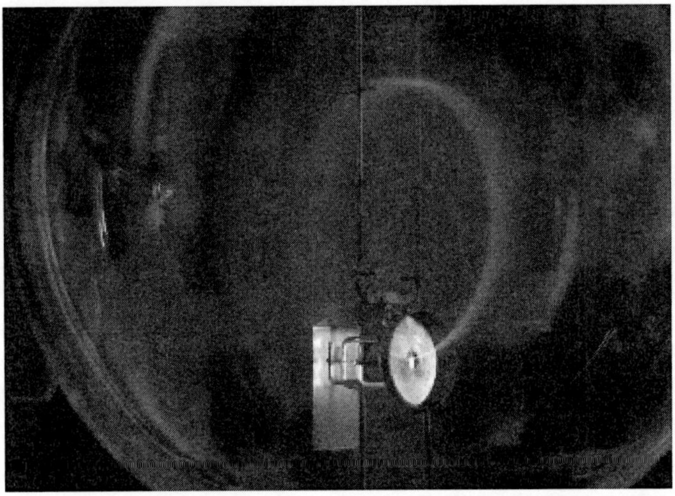

Figure: *Beam of electrons moving in a circle. Lighting is caused by excitation of gas atoms in a bulb.*

The electrodes shown at the right would be in the vacuum chamber, which is flat, in a narrow gap between the two poles of a large magnet. In the cyclotron, a high-frequency alternating voltage applied across the "D" electrodes (also called "dees") alternately attracts and repels charged particles.

The particles, injected near the centre of the magnetic field, increase in speed (and therefore energy) only when passing through the gap between the electrodes. The perpendicular magnetic field (passing vertically through the "D" electrodes), combined with the increasing energy of the particles forces the particles to travel in a spiral path. With no change in energy the charged particles in a magnetic field will follow a circular path.

In the cyclotron, energy is applied to the particles as they cross the gap between the dees and so they are accelerated (at the typical sub-relativistic speeds used) and will increase in mass as they approach the speed of light. Either of these effects (increased velocity or increased mass) will increase the radius of the circle and so the path will be a spiral. (The particles move in a spiral, because a current of electrons or ions, flowing perpendicular to a magnetic field, experiences a force perpendicular to its direction of motion. The charged particles move freely in a vacuum, so the particles follow a spiral path.) The radius will increase until the particles hit a target at the perimetre of the vacuum chamber. Various materials may be used for the target, and the collisions will create secondary particles which may be guided outside of the cyclotron and into instruments for analysis. The results will enable the calculation of various properties, such as the mean spacing between atoms and the creation of various collision products. Subsequent chemical and particle analysis of the target material may give insight into nuclear transmutation of the elements used in the target.

Uses of the Cyclotron

For several decades, cyclotrons were the best source of high-energy beams for nuclear physics experiments; several cyclotrons are still in use for this type of research. Cyclotrons can be used to treat cancer. Ion beams from cyclotrons can be used, as in proton therapy, to penetrate the body and kill tumours by radiation damage, while minimising damage to healthy tissue along their path. Cyclotron beams can be used to bombard other atoms to produce short-lived positron-emitting isotopes suitable for PET imaging.

Problems Solved by the Cyclotron

The cyclotron was an improvement over the linear accelerators that were available when it was invented. A linear accelerator (also called a linac) accelerates particles in a straight line through an evacuated tube (or series of such tubes placed end to end). A set of electrodes shaped like flat donuts are arranged inside the length of the tube(s).

These are driven by high-power radio waves that continuously switch between positive and negative voltage, causing particles travelling along the centre of the tube to accelerate.

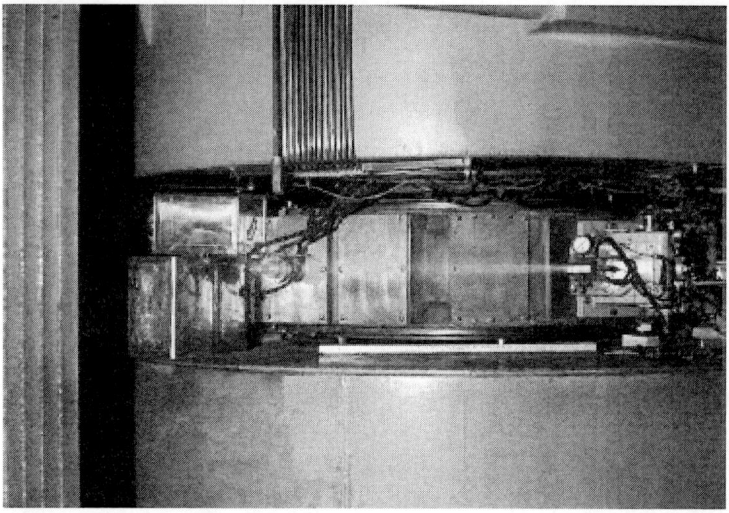

***Figure:** 60-inch cyclotron, circa 1939, showing a beam of accelerated ions (likely protons or deuterons) escaping the accelerator and ionizing the surrounding air causing a blue glow*

In the 1920s, it was not possible to get high frequency radio waves at high power, so either the accelerating electrodes had to be far apart to accommodate the low frequency or more stages were required to compensate for the low power at each stage. Either way, higher-energy particles required longer accelerators than scientists could afford. Modern linacs use high power Klystrons and other devices able to impart much more power at higher frequencies. But before these devices existed, cyclotrons were cheaper than linacs. Cyclotrons accelerate particles in a spiral path. Therefore, a compact accelerator can contain much more distance than a linear accelerator, with more opportunities to accelerate the particles.

Advantages of the Cyclotron

- Cyclotrons have a single electrical driver, which saves both money and power, since more expense may be allocated to increasing efficiency.
- Cyclotrons produce a continuous stream of particles at the target, so the average power is relatively high.
- The compactness of the device reduces other costs, such as its foundations, radiation shielding, and the enclosing building.

Limitations of the Cyclotron

The spiral path of the cyclotron beam can only "sync up" with klystron-type (constant frequency) voltage sources if the accelerated particles are approximately obeying Newton's Laws of Motion. If the particles become fast enough that relativistic effects become important, the beam gets out of phase with the oscillating electric field, and cannot receive any additional acceleration.

The cyclotron is therefore only capable of accelerating particles up to a few percent of the speed of light. To accommodate increased mass the magnetic field may be modified by appropriately shaping the pole pieces as in the isochronous cyclotrons, operating in a pulsed mode and changing the frequency applied to the dees as in the synchrocyclotrons, either of which is limited by the diminishing cost effectiveness of making larger machines. Cost limitations have been overcome by employing the more complex synchrotron or linear accelerator, both of which have the advantage of scalability, offering more power within an improved cost structure as the machines are made larger.

Mathematics of the Cyclotron

Non-relativistic

The centripetal force is provided by the transverse magnetic field *B*, and the force on a particle travelling in a magnetic field (which causes it to be angularly displaced, i.e. spiral) is equal to *Bqv*. So,

$$\frac{mv^2}{r} = Bqv$$

(Where m is the mass of the particle, q is its charge, B the magnetic field strength, v is its velocity and r is the radius of its path.)

The speed at which the particles enter the cyclotron due to a potential difference, V.

$$v = \sqrt{\frac{2Vq}{m}}$$

Therefore,

$$\frac{v}{r} = \frac{Bq}{m}$$

v/r is equal to angular velocity, ω, so

$$\omega = \frac{Bq}{m}$$

And since the angular frequency is

$$\omega = 2\pi f$$

Therefore,

$$f = \frac{Bq}{2\pi m}$$

The frequency of the driving voltage is simply the inverse of this frequency so that the particle crosses between the dees at the same point in the voltage cycle.

$$f_c = \frac{Bq}{2\pi m}$$

***Figure:** A pair of "dee" electrodes with loops of coolant pipes on their surface at the Lawrence Hall of Science. The particle exit point may be seen at the top of the upper dee, where the target would be positioned*

This shows that for a particle of constant mass, the frequency does not depend upon the radius of the particle's orbit.

As the beam spirals out, its frequency does not decrease, and it must continue to accelerate, as it is travelling more distance in the same time.

As particles approach the speed of light, they acquire additional mass, requiring modifications to the frequency, or the magnetic field during the acceleration.

This is accomplished in the synchrocyclotron.

Relativistic

The radius of curvature for a particle moving relativistically in a static magnetic field is

$$r = \frac{\gamma m v}{qB}$$

where

$$\gamma = \frac{1}{\sqrt{1-\left(\frac{v}{c}\right)^2}}$$ is the Lorentz factor

Note that in high-energy experiments energy, E, and momentum, p, are used rather than velocity, and both measured in units of energy. In that case one should use the substitution,

$$\frac{p}{E} = v$$

where this is in Natural units

The relativistic cyclotron frequency is

$$f = f_c\sqrt{1-\left(\frac{v}{c}\right)^2}$$

where f_c is the classical frequency, given above, of a charged particle with velocity v circling in a magnetic field.

The rest mass of an electron is 511 keV/c^2, so the frequency correction is 1% for a magnetic vacuum tube with a 5.11 keV/c^2 direct current accelerating voltage. The proton mass is nearly two thousand times the electron mass, so the 1% correction energy is about 9 MeV, which is sufficient to induce nuclear reactions.

An alternative to the synchrocyclotron is the isochronous cyclotron, which has a magnetic field that increases with radius, rather than with time. The de-focusing effect of this radial field gradient is compensated by ridges on the magnet faces which vary the field azimuthally as well. This allows particles to be accelerated continuously, on every period of the radio frequency (RF), rather than in bursts as in most other accelerator types. This principle that alternating field gradients have a net focusing effect is called strong focusing. It was obscurely known theoretically long before it was put into practice.

Related Technologies

- The spiralling of electrons in a cylindrical vacuum chamber within a transverse magnetic field is also employed in the magnetron, a device for producing high frequency radio waves (microwaves).

- The Synchrotron moves the particles through a path of constant radius, allowing it to be made as a pipe and so of much larger radius than is practical with the cyclotron and synchrocyclotron. The larger radius allows the use of numerous magnets, each of which imparts angular momentum and so allows particles of higher velocity (mass) to be kept within the bounds of the evacuated pipe. The magnetic field strength of each of the bending magnets is increased as the particles gain energy in order to keep the bending angle constant.

Gyrotron

Gyrotrons are high powered vacuum tubes which emitmillimetre-wave beams by bunching electrons with cyclotron motion in a strong magnetic field. Output frequencies range from about 20 to 250 GHz, covering wavelengths from microwave to the edge of the terahertzgap. Typical output powers range from tens of kilowatts to 1-2 megawatts. Gyrotrons can be designed for pulsed or continuous operation.

The sizes of the cavities determine the resonant frequency, and thereby the frequency of emitted microwaves. However, the frequency is not precisely controllable. The operating frequency varies with changes in load impedance, with changes in the supply current, and with the temperature of the tube.

This is not a problem in uses such as heating, or in some forms of radar where the receiver can be synchronized with an imprecise magnetron frequency. Where precise frequencies are needed, other devices such as the klystron are used.

The magnetron is a self-oscillating device requiring no external elements other than a power supply. A well-defined threshold anode voltage must be applied before oscillation will build up; this voltage is a function of the dimensions of the resonant cavity, and the applied magnetic field. In pulsed applications there is a delay of several cycles before the oscillator achieves full peak power, and the build-up of anode voltage must be coordinated with the build-up of oscillator output.

The magnetron is a fairly efficient device. In a microwave oven, for instance, a 1.1 kilowatt input will generally create about 700 watt of microwave power, an efficiency of around 65%. (The high-voltage and the properties of the cathode determine the power of a magnetron.)

Large S-band magnetrons can produce up to 2.5 megawatts peak power with an average power of 3.75 kW. Large magnetrons can be water cooled. The magnetron remains in widespread use in roles which require high power, but where precise frequency control is unimportant.

Applications

Radar: In radar devices, the waveguide is connected to an antenna. The magnetron is operated with very short pulses of applied voltage, resulting in a short pulse of high power microwave energy being radiated. As in all radar systems, the radiation reflected off a target is analysed to produce a radar map on a screen.

Several characteristics of the magnetron's power output conspire to make radar use of the device somewhat problematic. The first of these factors is the magnetron's inherent instability in its transmitter frequency. This instability is noted not only as a frequency shift from one pulse to the next, but also a frequency shift within an individual transmitter pulse. The second factor is that the energy of the transmitted pulse is spread over a wide frequency spectrum, which makes necessary its receiver to have a corresponding wide selectivity. This wide selectivity permits ambient electrical noise to be accepted into the receiver, thus obscuring somewhat the received radar echoes, thereby reducing overall radar performance. The third factor, depending on application, is the radiation hazard caused by the use of high power electromagnetic radiation. In some applications, for example a marine radar mounted on a recreational vessel, a radar with a magnetron output of 2 to 4 kilowatts is often found mounted very near an area occupied by crew or passengers. In practical use, these factors have been overcome, or merely accepted, and there are today thousands of magnetron aviation and marine radar units in service. Recent advances in aviation weather avoidance radar and in marine radar have successfully implemented semiconductor transmitters that eliminate the magnetron entirely.

Heating

In microwave ovens, the waveguide leads to a radio frequency-transparent port into the cooking chamber.

Lighting

In microwave-excited lighting systems, such as a sulfur lamp, a magnetron provides the microwave field that is passed through a waveguide to the lighting cavity containing the light-emitting substance (e.g., sulfur, metal halides, etc.)

History

The first simple, two-pole magnetron was developed in 1920 by Albert Hull at General Electric's Research Laboratories (Schenectady, New York), as an outgrowth of his work on the magnetic control of vacuum tubes in an attempt to work around the patents held by Lee De Forest on electrostatic control.

Hull's magnetron was not originally intended to generate VHF (very-high-frequency) electromagnetic waves. However, in 1924, Czech physicist August Žáθek (1886–1961) and German physicist Erich Habann (1892–1968) independently discovered that the magnetron could generate waves of 100 megahertz to 1 gigahertz. Žáθek, a professor at Prague's Charles University, published first; however, he published in a journal with a small circulation and thus attracted little attention. Habann, a student at the University of Jena, investigated the magnetron for his doctoral dissertation of 1924. Throughout the 1920s, Hull and other researchers around the world worked to develop the magnetron. Most of these early magnetrons were glass vacuum tubes with multiple anodes.

However, the two-pole magnetron, also known as a split-anode magnetron, had relatively low efficiency. The cavity version (properly referred to as a *resonant-cavity magnetron*) proved to be far more useful.

An early multi-cavity version of the magnetron was reported by Bucharest University Professor Theodor V. Ionescu. and followed in 1937-1940 by a similar multi-cavity magnetron built by the British physicist, Sir John Turton Randall, FRSE together with a team of British coworkers for the British and American, military radar installations in WWII.

While radar was being developed during World War II, there arose an urgent need for a high-power microwave generator that worked at shorter wavelengths (around 10 cm (3 GHz)) rather than the 150 cm (200 MHz) that was available from tube-based generators of the time. It was known that a multi-cavity resonant magnetron had been developed and patented in 1935 by Hans Hollmann inBerlin, and independently, in 1935, by physicist Theodor V. Ionescu in Romania. However, the German military considered the frequency drift of Hollman's device to be undesirable, and based their radar systems on the klystron instead. But klystrons could not achieve the high power output that magnetrons eventually reached. This was one reason that German night fighter radars were not a match for their British counterparts.

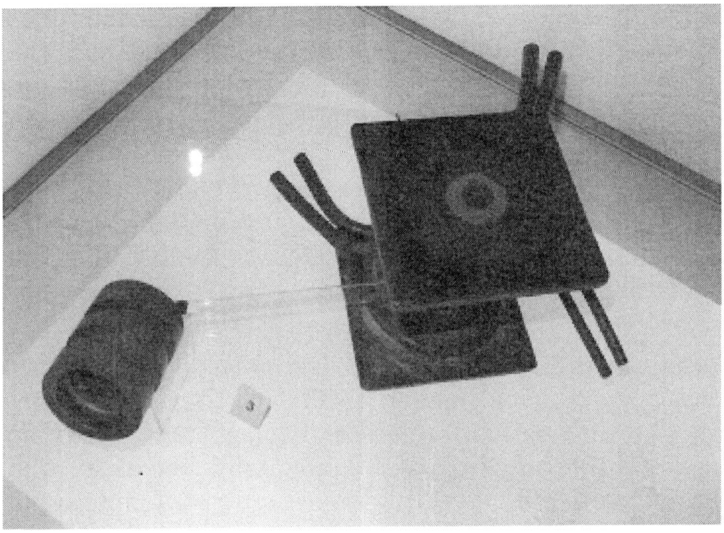

Figure: *Original Randall and Boot cavity magnetron*

In 1940, at the University of Birmingham in the United Kingdom, John Randall and Harry Boot produced a working prototype similar to Hollman's cavity magnetron, but added liquid cooling and a stronger cavity. Randall and Boot soon managed to increase its power output 100 fold. Instead of abandoning the magnetron due to its frequency instability, they sampled the output signal and synchronized their receiver to whatever frequency was actually being generated. In 1941, the problem of frequency instability was solved by coupling alternate cavities within the magnetron.

Because France had just fallen to the Nazis and Britain had no money to develop the magnetron on a massive scale, Churchill agreed that Sir Henry Tizard should offer the magnetron to the Americans in exchange for their financial and industrial help (the Tizard Mission). An early 6 kW version, built in England by the General Electric Company Research Laboratories, Wembley, London (not to be confused with the similarly named American company General Electric), was given to the US government in September 1940. At the time the most powerful equivalent microwave producer available in the US (a klystron) had a power of only ten watts. The cavity magnetron was widely used during World War II in microwave radar equipment and is often credited with giving Allied radar a considerable performance advantage over German and Japanese radars, thus directly influencing the outcome of the war. It was later described as "the most valuable cargo ever brought to our shores".

The Bell Telephone Laboratories made a producible version from the magnetron delivered to America by the Tizard Mission, and before the end of 1940, the Radiation Laboratory had been set up on the campus of the Massachusetts Institute of Technology to develop various types of radar using the magnetron. By early 1941, portable centimetric airborne radars were being tested in American and British planes. In late 1941, the Telecommunications Research Establishment in Great Britain used the magnetron to develop a revolutionary airborne, ground-mapping radar codenamed H2S. The H2S radar was in part developed by Alan Blumlein and Bernard Lovell.

Centimetric radar, made possible by the cavity magnetron, allowed for the detection of much smaller objects and the use of much smaller antennas. The combination of small-cavity magnetrons, small antennas, and high resolution allowed small, high quality radars to be installed in aircraft. They could be used by maritime patrol aircraft to detect objects as small as a submarine periscope, which allowed aircraft to attack and destroy submerged submarines which had previously been undetectable from the air. Centimetric contour mapping radars like H2S improved the accuracy of Allied bombers used in the strategic bombing campaign. Centimetric gun-laying radars were likewise far more accurate than the older technology. They made the big-gunned Allied battleships more deadly and, along with the newly developed proximity fuze, made anti-aircraft guns much more dangerous to attacking aircraft. The two coupled together and used by anti-aircraft batteries, placed along the flight path of German V-1 flying bombs on their way to London, are credited with destroying many of the flying bombs before they reached their target.

Since then, many millions of cavity magnetrons have been manufactured; while some have been for radar the vast majority have been for microwave ovens. The use in radar itself has dwindled to some extent, as more accurate signals have generally been needed and developers have moved to klystron and travelling-wave tube systems for these needs.

Health Hazards

Among more speculative hazards, at least one in particular is well known and documented. As the lens of the eye has no cooling blood flow, it is particularly prone to overheating when exposed to microwave radiation. This heating can in turn lead to a higher incidence of cataracts in later life. A microwave oven with a warped door or poor microwave sealing can be hazardous.

Figure: *Caution: radiowaves hazard*

There is also a considerable electrical hazard around magnetrons, as they require a high voltage power supply.

Some magnetrons have beryllium oxide (beryllia) ceramic insulators, which are dangerous if crushed and inhaled, or otherwise ingested. Single or chronic exposure can lead to berylliosis, an incurable lung condition.

In addition, beryllia is listed as a confirmed human carcinogen by the IARC; therefore, broken ceramic insulators or magnetrons should not be directly handled.

Klystron

A klystron is a specialised linear-beam vacuum tube(evacuated electron tube). Klystrons are used as amplifiers at microwave and radio frequencies to produce both low-power reference signals for superheterodyne radar receivers and to produce high-power carrier waves for communications and the driving force for modern particle accelerators.

Klystron amplifiers have the advantage (over the magnetron) of coherently amplifying a reference signal so its output may be precisely controlled in amplitude, frequency and phase. Many klystrons have a waveguide for coupling microwave energy into and out of the device, although it is also quite common for lower power and lower frequency klystrons to use coaxial couplings instead. In some cases a coupling probe is used to couple the microwave energy from a klystron into a separate external waveguide.

Figure: *High-power klystron used for spacecraft communication at the Canberra Deep Space Communications Complex.*

All modern klystrons are amplifiers, since reflex klystrons, which were used as oscillators in the past, have been surpassed by alternative technologies. The name *klystron* comes from the stem form κλυό- (*klys*) of a Greek verb referring to the action of waves breaking against a shore, and the end of the word *electron*.

History

The brothers Russell and Sigurd Varian of Stanford University are the inventors of the klystron. Their prototype was completed in August 1937. Upon publication in 1939, news of the klystron immediately influenced the work of US and UK researchers working on radar equipment. The Varians went on to found Varian Associates to commercialise the technology (for example to make small linear accelerators to generate photons for external beam radiation therapy).

In their 1939 paper, they acknowledged the contribution of A. Arsenjewa-Heil and Oskar Heil (wife and husband) for their velocity modulation theory in 1935.

The work of physicist W.W. Hansen was instrumental in the development of the klystron and was cited by the Varian brothers in their 1939 paper. His resonator analysis, which dealt with the problem of accelerating electrons toward a target, could be used just as well to decelerate electrons (i.e., transfer their kinetic energy to RF energy in a resonator). During the second World War, Hansen lectured at the MIT Radiation labs two days a week, commuting to Boston from Sperry gyroscope company on Long Island. His resonator was called a "rhumbatron" by the Varian brothers. Hansen died of beryllium disease in 1949 as a result of exposure to beryllium oxide (BeO).

During the second World War, the Axis powers relied mostly on (then low-powered and long wavelength) klystron technology for their radar system microwave generation, while the Allies used the far more powerful but frequency-drifting technology of the Cavity magnetron for much shorter one centimetre microwave generation. Klystron tube technologies for very high-power applications, such as synchrotrons and radar systems, have since been developed.

Right after WW II AT&T used 4 watt klystrons in its brand new network of microwave links that covered the US continent. The network provided long distance telephone service and also carried television signals for the major TV networks.

Explanation

Klystrons amplify RF signals by converting the kinetic energy in a DC electron beam into radio frequency power. A beam of electrons is produced by a thermionic cathode (a heated pellet of lowwork function material), and accelerated by high-voltage electrodes (typically in the tens of kilovolts).

This beam is then passed through an input cavity. RF energy is fed into the input cavity at, or near, its natural frequency to produce a voltage which acts on the electron beam. The electric field causes the electrons to bunch: electrons that pass through during an opposing electric field are accelerated and later electrons are slowed, causing the previously continuous electron beam to form bunches at the input frequency.

To reinforce the bunching, a klystron may contain additional "buncher" cavities. The RF current carried by the beam will produce an RF magnetic field, and this will in turn excite a voltage across the gap of subsequent resonant cavities. In the output cavity, the developed RF energy is coupled out. The spent electron beam, with reduced energy, is captured in a collector.

Two-Cavity Klystron Amplifier

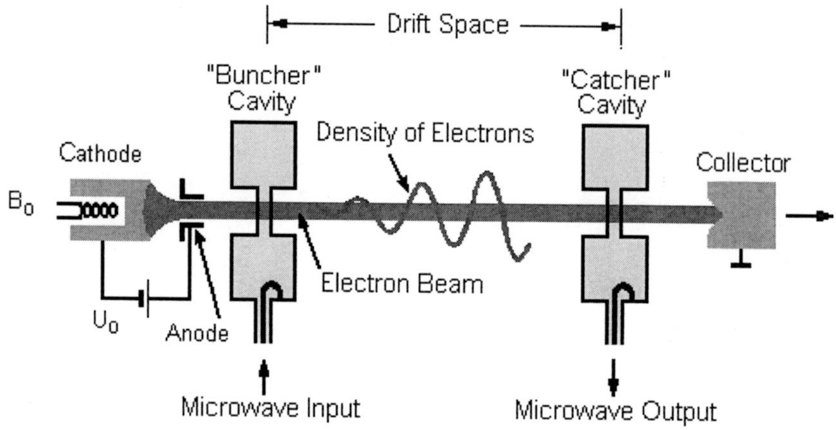

In the two-chamber klystron, the electron beamis injected into a resonant cavity. The electron beam, accelerated by a positive potential, is constrained to travel through a cylindrical *drift tube* in a straight path by an axial magnetic field.

While passing through the first cavity, the electron beam is velocity modulated by the weak RF signal. In the moving frame of the electron beam, the velocity modulation is equivalent to a plasma oscillation. Plasma oscillations are rapid oscillations of the electron density in conducting media such as plasmas or metals.

(The frequency only depends weakly on the wavelength). So in a quarter of one period of the plasma frequency, the velocity modulation is converted to density modulation, i.e. bunches of electrons.

As the bunched electrons enter the second chamber they inducc standing waves at the same frequency as the input signal. The signal induced in the second chamber is much stronger than that in the first.

Two-Cavity Klystron Oscillator

The two-cavity amplifier klystron is readily turned into an oscillator klystron by providing a feedbackloop between the input and output cavities.

Two-cavity oscillator klystrons have the advantage of being among the lowest-noise microwave sources available, and for that reason have often been used in the illuminator systems of missile targeting radars.

The two-cavity oscillator klystron normally generates more power than the reflex klystron—typically watts of output rather than milliwatts. Since there is no reflector, only one high-voltage supply is necessary to cause the tube to oscillate, the voltage must be adjusted to a particular value.

This is because the electron beam must produce the bunched electrons in the second cavity in order to generate output power. Voltage must be adjusted to vary the velocity of the electron beam (and thus the frequency) to a suitable level due to the fixed physical separation between the two cavities.

Often several "modes" of oscillation can be observed in a given klystron. In the reflex klystron (also known as a 'Sutton' klystron after its inventor, Robert Sutton), the electron beam passes through a single resonant cavity.

Reflex Klystron

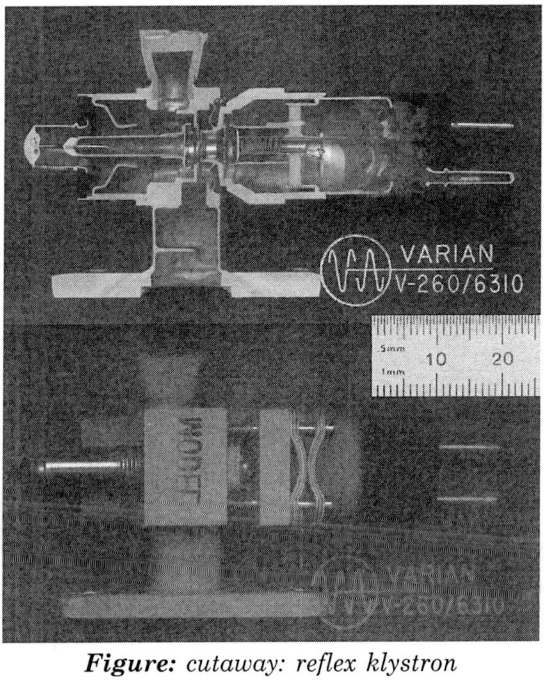

***Figure:** cutaway: reflex klystron*

The electrons are fired into one end of the tube by an electron gun. After passing through the resonant cavity they are reflected by a negatively charged reflector electrode for another pass through the cavity, where they are then collected. The electron beam is velocity modulated when it first passes through the cavity. The formation of electron bunches takes place in the drift space between the reflector and the cavity.

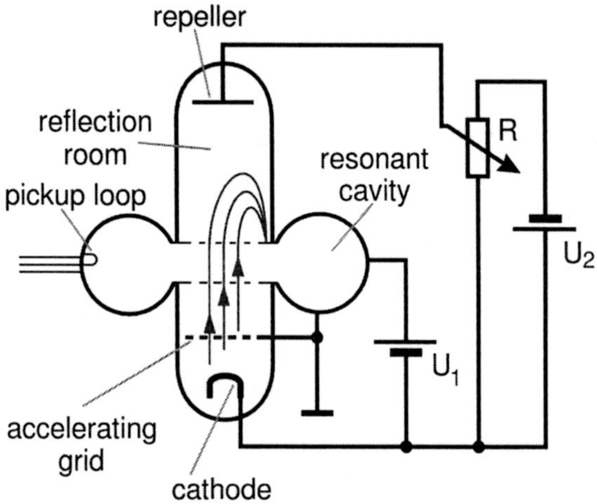

The voltage on the reflector must be adjusted so that the bunching is at a maximum as the electron beam re-enters the resonant cavity, thus ensuring a maximum of energy is transferred from the electron beam to the RF oscillations in the cavity. The voltage should always be switched on before providing the input to the reflex klystron as the whole function of the reflex klystron would be destroyed if the supply is provided after the input. The reflector voltage may be varied slightly from the optimum value, which results in some loss of output power, but also in a variation in frequency. This effect is used to good advantage for automatic frequency control in receivers, and in frequency modulation for transmitters. The level of modulation applied for transmission is small enough that the power output essentially remains constant. At regions far from the optimum voltage, no oscillations are obtained at all. This tube is called a reflex klystron because it repels the input supply or performs the opposite function of a klystron.

There are often several regions of reflector voltage where the reflex klystron will oscillate; these are referred to as modes. The electronic tuning range of the reflex klystron is usually referred to as the variation in frequency between half power points—the points in the oscillating mode where the power output is half the maximum output in the mode. The frequency of oscillation is dependent on the reflector voltage, and varying this provides a crude method of frequency modulating the oscillation frequency, albeit with accompanying amplitude modulation as well.

Modern semiconductor technology has effectively replaced the reflex klystron in most applications.

Multicavity Klystron

In all modern klystrons, the number of cavities exceeds two. A larger number of cavities may be used to increase the gain of the klystron, or to increase the bandwidth.

Tuning a Klystron

Some klystrons have cavities that are tunable. Tuning a klystron is delicate work which, if not done properly, can cause damage to equipment or injury to the technician. By adjusting the frequency of individual cavities, the technician can change the operating frequency, gain, output power, or bandwidth of the amplifier. The technician must be careful not to exceed the limits of the graduations, or damage to the klystron can result.

Manufacturers generally send a card with the unique calibrations for a klystron's performance characteristics, that lists the graduations to be set to attain any of a set of listed frequencies. No two klystrons are exactly identical (even when comparing like part/model number klystrons), and so every card is specific to the individual unit. Klystrons have serial numbers on each of them to uniquely identify each unit, and for which manufacturers may (hopefully) have the performance characteristics in a database. If not, loss of the calibration card may be an economically insoluble problem, making the klystron unusable or perform marginally un-tuned.

Other precautions taken when tuning a klystron include using nonferrous tools. Some klystrons employ permanent magnets. If a technician uses ferrous tools, (which are ferromagnetic), and comes too close to the intense magnetic fields that contain the electron beam, such a tool can be pulled into the unit by the intense magnetic force, smashing fingers, injuring the technician, or damaging the unit. Special lightweight nonmagnetic (aka diamagnetic) tools made of beryllium alloy have been used for tuning U.S. Air Force klystrons.

Precautions are routinely taken when transporting klystron devices in aircraft, as the intense magnetic field can interfere with magnetic navigation equipment. Special overpacks are designed to help limit this field "in the field," and thus allow such devices to be transported safely.

Optical Klystron

In an optical klystron the cavities are replaced with undulators. Very high voltages are needed. The electron gun, the drift tube and the collector are still used.

Floating Drift Tube Klystron

The floating drift tube klystron has a single cylindrical chamber containing an electrically isolated central tube. Electrically, this is similar to the two cavity oscillator klystron with a lot of feedback between the two cavities. Electrons exiting the source cavity are velocity modulated by the electric field as they travel through the drift tube and emerge at the destination chamber in bunches, delivering power to the oscillation in the cavity. This type of oscillator klystron has an advantage over the two-cavity klystron on which it is based. It only needs one tuning element to effect changes in frequency. The drift tube is electrically insulated from the cavity walls, and DC bias is applied separately. The DC bias on the drift tube may be adjusted to alter the transit time through it, thus allowing some electronic tuning of the oscillating frequency. The amount of tuning in this manner is not large and is normally used for frequency modulation when transmitting.

Collector

After the RF energy has been extracted from the electron beam, the beam is destroyed in a collector. Some klystrons include depressed collectors, which recover energy from the beam before collecting the electrons, increasing efficiency. Multistage depressed collectors enhance the energy recovery by "sorting" the electrons in energy bins.

Applications

Klystrons produce microwave power far in excess of that developed by solid state. In modern systems, they are used from UHF (hundreds of MHz) up through hundreds of gigahertz (as in the Extended Interaction Klystrons in the CloudSat satellite). Klystrons can be found at work in radar, satellite and wideband high-power communication (very common in television broadcasting and EHF satelliteterminals), medicine (radiation oncology), and high-energy physics (particle accelerators and experimental reactors). At SLAC, for example, klystrons are routinely employed which have outputs in the range of 50 megawatts (pulse) and 50 kilowatts (time-averaged) at frequencies nearing 3 GHz.

Popular Science's "Best of What's New 2007" described a company, Global Resource Corporation, using a klystron to convert the hydrocarbons in everyday materials, automotive waste, coal, oil shale, and oil sands into natural gas and diesel fuel.

Travelling-Wave Tube

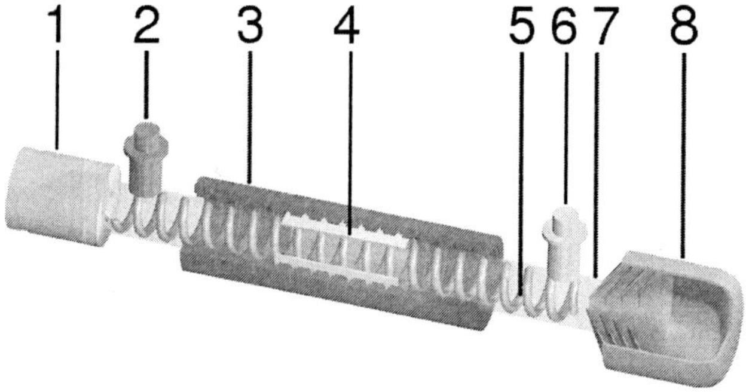

Figure: *Cutaway view of a helix TWT. (1) Electron gun; (2) RF input; (3) Magnets; (4) Attenuator; (5) Helix coil; (6) RF output; (7) Vacuum tube; (8) Collector.*

A travelling-wave tube (TWT) is an electronic device used to amplify radio frequency (RF) signals to high power, usually in an electronic assembly known as a travelling-wave tube amplifier (TWTA).

The bandwidth of a broadband TWT can be as high as one octave, although tuned (narrowband) versions exist, and operating frequencies range from 300 MHz to 50 GHz. The voltage gain of the tube can be of the order of 70 decibels.

Description

The device is an elongated vacuum tube with an electron gun (a heated cathode that emits electrons) at one end. A magnetic containment field around the tube focuses the electrons into a beam, which then passes down the middle of an RF circuit (wire helix or coupled cavity) that stretches from the RF input to the RF output, the electron beam finally striking a collector at the other end. A directional coupler, which can be either a waveguide or an electromagnetic coil, fed with the low-powered radio signal that is to be amplified, is positioned near the emitter, and induces a current into the helix. The RF circuit acts as a delay line, in which the RF signal travels at near the same speed along the tube as the electron beam.

The electromagnetic field due to the RF signal in the RF circuit interacts with the electron beam, causing bunching of the electrons (an effect called *velocity modulation*), and the electromagnetic field due to the beam current then induces more current back into the RF circuit (i.e. the current builds up and thus is amplified as it passes down).

A second directional coupler, positioned near the collector, receives an amplified version of the input signal from the far end of the RF circuit. Attenuator(s) placed along the RF circuit prevents reflected wave from travelling back to the cathode.

Higher powered Helix TWTs usually contain beryllium oxide ceramic as both a helix support rod and in some cases, as an electron collector for the TWT because of its special electrical, mechanical, and thermal properties.

Invention, Development and Early Use

The invention of the TWT is widely attributed to Rudolf Kompfner in 1942–1943, though Nils Lindenblad working at RCA (Radio Corporation of America) in the USA did patent a device in May 1940 that was remarkably similar to Kompfner's TWT. Kompfner independently invented the TWT, and built the first working TWT, in a British Admiralty radar lab during World War II.

His first sketch of his TWT is dated November 12, 1942, and he built his first TWT in early 1943. The TWT was refined by Kompfner, John Pierce, and Lester M. Field at Bell Labs.

By the fifties, after further development at the Electron Tube Laboratory at Hughes Aircraft Company in Culver City, California, TWTs went into production there, and by the sixties TWTs were also produced by such companies as the English Electric Valve Company, followed by Ferranti in the seventies.

On July 10, 1962, the first communications satellite, Telstar 1, was launched with a 2 W, 4 GHz RCA-designed TWT transponder used for transmitting RF signals back to the earth. Syncom 2, the first synchronous satellite (Syncom 1 did not reach its final orbit), launched on July 26, 1963 with two 2 W, 1850 MHz Hughes-designed TWT transponders (one active and one spare).

Coupled-Cavity TWT

Helix TWTs are limited in peak RF power by the current handling (and therefore thickness) of the helix wire. As power level increases, the wire can overheat and cause the helix geometry to warp. Wire thickness can be increased to improve matters, but if the wire is too thick it becomes impossible to obtain the required helix pitch for proper operation. Typically helix TWTs achieve less than 2.5 kW output power.

The coupled-cavity TWT overcomes this limit by replacing the helix with a series of coupled cavities arranged axially along the beam. Conceptually, this structure provides a helical waveguide and hence amplification can occur via velocity modulation. Helical waveguides have very nonlinear dispersion and thus are only narrowband (but wider than klystron). A coupled-cavity TWT can achieve 60 kW output power.

Operation is similar to that of a klystron, except that coupled-cavity TWTs are designed with attenuation between the slow-wave structure instead of a drift tube. The slow-wave structure gives the TWT its wide bandwidth. A free electron laser allows higher frequencies.

Travelling-wave-tube Amplifier

A TWT integrated with a regulated power supply and protection circuits is referred to as a travelling-wave-tube amplifier (abbreviated TWTA and often pronounced "TWEET-uh"). It is used to produce high-power radio frequency signals. The bandwidth of a broadband TWTA can be as high as oneoctave, although tuned (narrowband) versions exist; operating frequencies range from 300 MHz to 50 GHz.

A TWTA consists of a travelling-wave tube coupled with its protection circuits (as in klystron) and regulated power supply Electronic Power Conditioner (EPC), which may be supplied and integrated by a different manufacturer.

The main difference between most power supplies and those for vacuum tubes is that efficient vacuum tubes have depressed collectors to recycle kinetic energy of the electrons and therefore the secondary winding of the power supply needs up to 6 taps of which the helix voltage needs precise regulation.

The subsequent addition of a linearizer (as for inductive output tube) can, by complementary compensation, improve the gain compression and other characteristics of the TWTA; this combination is called a linearized TWTA (LTWTA, "EL-tweet-uh"). Broadband TWTAs generally use a helix TWT, and achieve less than 2.5 kW output power. TWTAs using a coupled cavity TWT can achieve 15 kW output power, but at the expense of bandwidth.

Uses

TWTAs are commonly used as amplifiers in satellite transponders, where the input signal is very weak and the output needs to be high power. A TWTA whose output drives an antenna is a type of transmitter. TWTA transmitters are used extensively in radar, particularly in airborne fire-control radar systems, and in electronic warfare and self-protection systems. In these types of applications, a control grid is typically introduced between the TWT's electron gun and slow-wave structure to allow pulsed operation. The circuit that drives the control grid is usually referred to as a grid modulator. Another major use of TWTAs is for the electromagnetic compatibility (EMC) testing industry for immunity testing of electronic devices.

Crossed-Field Amplifier

A crossed-field amplifier (CFA) is a specialised vacuum tube, first introduced in the mid-1950s and frequently used as a microwave amplifier in very-high-power transmitters. Raytheon engineer William C. Brown's work to adapt magnetron principles to create a new broadband amplifier is generally recognised as the first CFA, which he called an Amplitron. Other names that are sometimes used by CFA manufacturers include Platinotron or Stabilotron.

A CFA has lower gain and bandwidth than other microwave amplifier tubes (such as klystrons or travelling-wave tubes); but it is more efficient and capable of much higher outputpower.

Peak output powers of many megawatts and average power levels of tens of kilowatts can be achieved, with efficiency ratings in excess of 70 percent. Their current use is in ground stations for TVRO broadcasting and Deep Space telecommunications networks.

Operation

The electric and magnetic fields in a CFA are perpendicular to each other ("crossed fields"). This is the same type of field interaction used in a magnetron; as a result, the two devices share many characteristics (such as high peak power and efficiency) and they have similar physical appearances. However, a magnetron is an oscillator and a CFA is an amplifier; a CFA's RF circuit (or slow-wave structure) is similar to that in a coupled-cavity TWT.

The CFA has the useful property that when power is shut off, the input simply passes to the output with very little loss. This avoids the need for RF bypass switching in the event of failure.

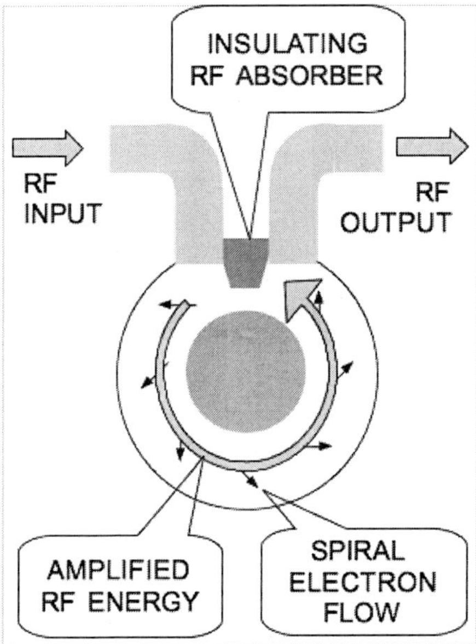

Figure: *Crossed-field amplifier internal operation.*

Two redundant CFAs can be connected sequentially, with only one powered; if it fails, power can be removed from the primary tube and applied to the secondary as a backup.

This approach was used on the S-band downlink transmitter on the Apollo Lunar Module where high efficiency and reliability were needed.

A large negative voltage is placed on the green electrode in the centre, and a large magnetic filed is directed perpendicular to the page. This forms a thin spinning disk of electrons with a flow pattern like spinning water as it drains from a sink or toilet. A slow-wave structure is located above and below the spinning disk of electrons. Electrons flow much slower than the speed of light, and the slow wave structure reduces the velocity of the input RF enough to match the electron velocity.

The RF input is introduced into the slow wave structure. The alternating microwave field causes the electrons to alternately speed up and slow down. These disturbances grow larger as electrons spiral around the device, and electrons slow down as the RF energy grows. This produces amplification.

There is a small amount of RF feedback from output to input. This creates a slight random phase jitter when the device is pulsed.

Backward Wave Oscillator

A backward wave oscillator (BWO), also called carcinotron (a trade name for tubes manufactured by CSF, now Thales) or backward wave tube, is a vacuum tube that is used to generate microwaves up to the terahertz range. It belongs to the travelling-wave tube family. It is an oscillator with a wide electronic tuning range.

An electron gun generates an electron beam that is interacting with a slow-wave structure. It sustains the oscillations by propagating a travelling wave backwards against the beam.

The generated electromagnetic wavepower has its group velocity directed oppositely to the direction of motion of the electrons. The output power is coupled out near the electron gun.

It has two main subtypes, the M-type, the most powerful, (M-BWO) and the O-type (O-BWO). The O-type delivers typically power in the range of 1 mW at 1000 GHz to 50 mW at 200 GHz. Carcinotrons are used as powerful and stable microwave sources. Due to the good quality wavefront they produce, they find use as illuminators in terahertz imaging.

The backward wave oscillators were demonstrated in 1951, M-type by Bernard Epsztein, (French patent 1,035,379; British patent 699,893; US patent 2,880,355) and O-type by Rudolf Kompfner.

The M-type BWO is a voltage-controlled non-resonant extrapolation of magnetron interaction, both types are tunable over a wide range of frequencies by varying the accelerating voltage.

They can be swept through the band fast enough to be appearing to radiate over all the band at once, which makes them suitable for effective radar jamming, quickly tuning into the radar frequency.

Carcinotrons allowed airborne radar jammers to be highly effective. However, frequency-agile radars can hop frequencies fast enough to force the jammer to use barrage jamming, diluting its output power over a wide band and significantly impairing its efficiency.

Carcinotrons are used in research, civilian and military applications.

For example, the Kopac passive sensor and Ramona passive sensor employed carcinotrons in their receiver systems.

The Slow-Wave Structure

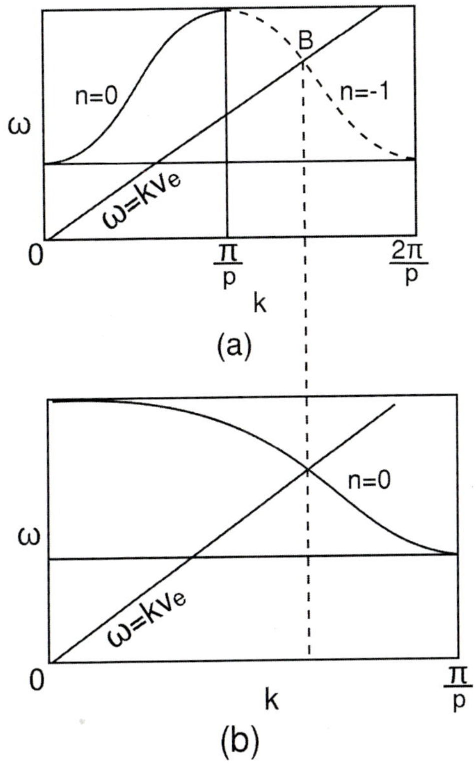

Figure: *(a) Forward fundamental space harmonic (n=0), (b) Backward fundamental*

The needed slow-wave structures must support a Radio Frequency (RF) electric field with a longitudinal component; the structures are periodic in the direction of the beam and behave like microwave filters with passbands and stopbands. Due to the periodicity of the geometry, the fields are identical from cell to cell except for a constant phase shift Φ.

This phase shift, a purely real number in a passband of a lossless structure, varies with frequency. According to Floquet's theorem, the RF electric field E(z,t) can be described at an angular frequency ω, by a sum of an infinity of "spatial or space harmonics" E_n

$$E(z,t)=\sum_{n=-\infty}^{+\infty} E_n e^{j(\omega t-k_n z)}$$

where the wave number or propagation constant k_n of each harmonic is expressed as:

$$k_n=(\Phi+2n\pi)/p \quad (-\pi<\Phi<+\text{ï})$$

z being the direction of propagation, p the pitch of the circuit and n an integer.

Two examples of slow-wave circuit characteristics are shown, in the ω-k or Brillouin diagram:

- on figure (a), the fundamental n=0 is a forward space harmonic (the phase velocity $v_n=\omega/k_n$ has the same sign as the group velocity $v_g=d\omega/dk_n$), synchronism condition for backward interaction is at point B, intersection of the line of slope v_e - the beam velocity - with the first backward (n = -1) space harmonic,
- on figure (b) the fundamental (n=0) is backward

A periodic structure can support both forward and backward space harmonics, which are not modes of the field, and cannot exist independently, even if a beam can be coupled to only one of them.

As the magnitude of the space harmonics decreases rapidly when the value of n is large, the interaction can be significant only with the fundamental or the first space harmonic.

M-type BWO

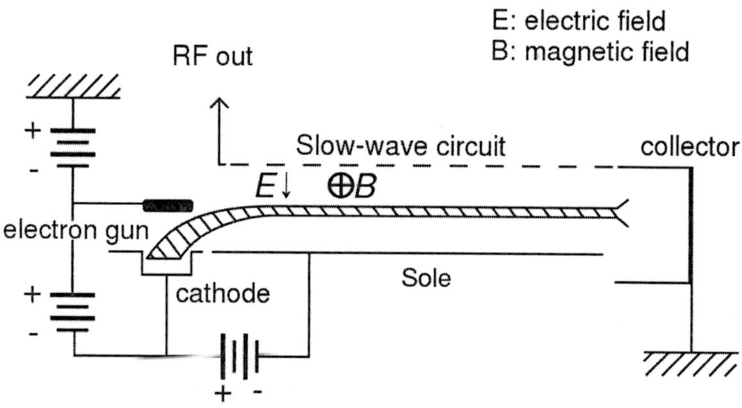

Figure: *Schematic of an M-BWO*

The M-type carcinotron, or M-type backward wave oscillator, uses crossed static electric field E and magnetic field B, similar to the magnetron, for focussing an electron sheet beam drifting perpendicularly to E and B, along a slow-wave circuit, with a velocity

E/B. Strong interaction occurs when the phase velocity of one space harmonic of the wave is equal to the electron velocity. Both E_z and E_y components of the RF field are involved in the interaction (E_y parallel to the static E field). Electrons which are in a decelerating E_z electric field of the slow-wave, lose the potential energy they have in the static electric field E and reach the circuit. The sole electrode is more negative than the cathode, in order to avoid collecting those electrons having gained energy while interacting with the slow-wave space harmonic.

O-type BWO

The O-type carcinotron, or O-type backward wave oscillator, uses an electron beam longitudinally focused by a magnetic field, and a slow-wave circuit interacting with the beam. A collector collects the beam at the end of the tube.

O-BWO Spectral Purity and Noise

1-Frequency Stability: The frequency-voltage sensitivity, is given by the relation

$$\Delta f/f=1/2[1/(1+|v_\Phi/v_g|)](\Delta V_0/V_0)$$

The oscillation frequency is also sensitive to the beam current (called "frequency pushing"). The current fluctuations at low frequencies are mainly due to the anode voltage supply, and the sensitivity to the anode voltage is given by

$$\Delta f/f=3/4[\omega_q/\omega/(1+|v_\Phi/v_g|)](\Delta V_a/V_a)$$

This sensitivity as compared to the cathode voltage sensitivity, is reduced by the ratio ω_q/ω, where ω_q i s the angular plasma frequency; this ratio is of the order of a few times 10^{-2}.

2-Noise

Measurements on submillimetre-wave BWO's (de Graauw et al., 1978) have shown that a signal-noise ratio of 120 dB per MHz could be expected in this wavelength range. In heterodyne detection using a BWO as a local oscillator, this figure corresponds to a noise temperature added by the oscillator of only 1000-3000K.

Free-Electron Laser

A free-electron laser, or FEL, is a laser that shares the same optical properties as conventional lasers such as emitting a beam consisting of coherent electromagnetic radiation which can reach high power, but which uses some very different operating principles to form the beam.

Unlike gas, liquid, or solid-state lasers such as diode lasers, in which electrons are excited in bound atomic or molecular states, FELs use a relativistic electron beam as the lasing medium which moves freely through a magnetic structure, hence the term *free electron.*

Figure: *Free electron laser schematic of operation*

The free-electron laser has the widest frequency range of any laser type, and can be widely tunable, currently ranging in wavelength from microwaves, through terahertz radiation and infrared, to the visible spectrum, to ultraviolet, to X-rays.

Free-electron lasers were invented by John Madey in 1976 at Stanford University. The work emanates from research done by Hans Motz who proposed the wiggler magnetic configuration which is at the heart of a free electron laser.

Madey used a 24 MeV electron beam and 5 m long wiggler to amplify a signal. Soon afterward, other laboratories with accelerators started developing such lasers.

Beam Creation

Figure: *Undulator of FELIX*

To create a FEL, a beam of electrons is accelerated to almost light speed. The beam passes through the FEL oscillator, a periodic transverse magnetic field produced by an arrangement of magnets with alternating poles within a laser cavity along the beam path. This array of magnets is sometimes called an undulator, or a "wiggler", because it forces the electrons in the beam to follow a sinusoidal path. The acceleration of the electrons along this path results in the release of photons (synchrotron radiation). Since the electron motion is in phase with the field of the light already emitted, the fields add together coherently. Whereas conventional undulators would cause the electrons to radiate independently, instabilities in the electron beam resulting from the interactions of the oscillations of electrons in the undulators and the radiation they emit leads to a bunching of the electrons, which continue to radiate in phase with each other. The wavelength of the light emitted can be readily tuned by adjusting the energy of the electron beam or the magnetic field strength of the undulators..

Accelerators

Today, a free-electron laser requires the use of an electron accelerator with its associated shielding, as accelerated electrons are a radiation hazard. These accelerators are typically powered by klystrons, which require a high voltage supply. The electron beam must be maintained in a vacuum which requires the use of numerous vacuum pumps along the beam path. While this equipment is bulky and expensive, free-electron lasers can achieve very high peak powers, and the tunability of FELs makes them highly desirable in several disciplines, including medical diagnosis and non-destructive testing.

X-ray uses

The lack of suitable mirrors in the extreme ultraviolet and x-ray regimes prevents the operation of a FEL oscillator; consequently, there must be suitable amplification over a single pass of the electron beam through the undulator to make the FEL worthwhile. X-ray free electron lasers use long undulators.. The underlying principle of the intense pulses from the X-ray laser lies in the principle of self-amplified stimulated emission (SASE), which leads to the microbunching of the electrons. Initially all electrons are distributed evenly and they emit incoherent spontaneous radiation only. Through the interaction of this radiation and the electrons' oscillations, they drift into microbunches separated by a distance equal to one radiation wavelength. Through this interaction, all electrons begin emitting coherent radiation in phase. In other words, all emitted radiation can reinforce itself perfectly whereby wave crests and wave troughs are always superimposed on one another in the best possible way. This results in an exponential increase of emitted radiation power, leading to high beam intensities and laser-like properties. Examples of facilities operating on the SASE FEL principle include the Free electron LASer (FLASH) in Hamburg, the Linac Coherent Light Source (LCLS) at the SLAC National Accelerator Laboratory, the European x-ray free electron laser (XFEL) in Hamburg, the SPring-8Compact SASE Source (SCSS), the SwissFEL at the Paul Scherrer Institute (Switzerland) and, as of 2011, the SACLA at the RIKEN Harima Institute in Japan.

One problem with SASE FELs is the lack of temporal coherence due to a noisy startup process. To avoid this, one can "seed" an FEL with a laser tuned to the resonance of the FEL. Such a temporally coherent seed can be produced by more conventional means, such as by high-harmonic generation(HHG) using an optical laser pulse. This results in coherent amplification of the input signal; in effect, the output laser quality is characterized by the seed. While HHG seeds are available at wavelengths down to the extreme ultraviolet, seeding is not feasible at x-ray wavelengths due to the lack of conventional x-ray lasers. FERMI@Elettra is a single-pass FEL user-facility covering the wavelength range from 100 nm (12 eV) to 10 nm (124 eV), located next to the third-generation synchrotron radiation facility ELETTRA in Trieste, Italy. The advent of femtosecond lasers has revolutionized many areas of science from solid state physics to biology. This new research frontier of ultra-fast VUV and X-ray science drives the development of a novel source for the generation of femtosecond pulses.

Medical Uses

Research by Dr. Glenn Edwards and colleagues at Vanderbilt University's FEL Centre in 1994 found that soft tissues like skin, cornea, and brain tissue could be cut, or ablated, using infrared FEL wavelengths around 6.45 micrometres with minimal collateral damage to adjacent tissue. This led to further research and eventually surgeries on humans, the first ever using a free-electron laser. Starting in 1999, and using the Keck foundation funded FEL operating rooms at the Vanderbilt FEL Centre, Dr. Michael Copeland and Dr. Pete Konrad of Vanderbilt performed three surgeries in which they resected meningioma brain tumors. Beginning in 2000, Dr. Karen Joos and Dr. Louise Mawn performed five surgeries involving the cutting of a window in the sheath of the optic nerve, to test the efficacy for optic nerve sheath fenestration. These eight surgeries went as expected with results consistent with the routine standard of care and with the added benefit of laser surgery and minimal collateral damage. A review of FELs for medical uses is given in the 1st edition of Tunable Laser Applications.

Since these successful results, there have been several efforts to build small, clinical lasers tunable in the 6 to 7 micrometre range with pulse structure and energy to give minimal collateral damage in soft tissue. At Vanderbilt, there exists a Raman shifted system pumped by an Alexandrite laser.

At the 2006 annual meeting of the American Society for Laser Medicine and Surgery (ASLMS), Dr. Rox Anderson of the Wellman Laboratory of Photomedicine of Harvard Medical School and Massachusetts General Hospital reported on the possible medical application of the free-electron laser in melting fats without harming the overlying skin. It was reported that at infrared wavelengths, water in tissue was heated by the laser, but at wavelengths corresponding to 915, 1210 and 1720 nm, subsurface lipids were differentially heated more strongly than water.

The possible applications of this selective photothermolysis (heating tissues using light) include the selective destruction of sebum lipids to treat acne, as well as targeting other lipids associated with cellulite and body fat as well as fatty plaques that form in arteries which can help treat atherosclerosis and heart disease.

Military Uses

FEL technology is being evaluated by the US Navy as a good candidate for an anti aircraft and missiledirected-energy weapon. Significant progress is being made in raising FEL power levels (the Thomas Jefferson National Accelerator Facility's FEL has demonstrated over 14 kW) and it should be possible to build compact multi-megawatt class FEL weapons.

On June 9, 2009 the Office of Naval Research announced it had awarded Raytheon a contract to develop a 100 kW experimental FEL. On March 18, 2010 Boeing Directed Energy Systems announced the completion of an initial design for U.S. Naval use. A prototype FEL system has since been demonstrated with a full-power prototype scheduled by 2018.

Maser

A maser is a device that produces coherentelectromagnetic waves through amplification by stimulated emission. Historically, "maser" derives from the original, upper-case acronym MASER, which stands for "Microwave Amplification by stimulated Emission of Radiation". The lower-case usage arose from technological development having rendered the original denotation imprecise, because contemporary masers emit EM waves (microwave and radio frequencies) across a broader band of the electromagnetic spectrum; thus, the physicist Charles H. Townes's suggested usage of "molecular" replacing "microwave", for contemporary linguistic accuracy. In 1957, when the optical coherent oscillator was first developed, it was denominated *optical maser*, but usually called laser (Light Amplification by Stimulated Emission of Radiation), the acronym Gordon Gould established in 1957.

History

Theoretically, the principle of the maser was described by Nikolay Basov and Alexander Prokhorov from Lebedev Institute of Physics at an *All-Union Conference on Radio-Spectroscopy* held by USSR Academy of Sciences in May 1952.

The results were subsequently published in October 1954. A precursor of the maser was the first *show boosted* hydrogen device built and tested by physicists Theodor V. Ionescu and Vasile Mihu in 1946. Independently, Charles H. Townes, J. P. Gordon, and H. J. Zeiger built the first ammonia maser at Columbia University in 1953. The device used stimulated emission in a stream of energized ammonia molecules to produce amplification of microwaves at a frequency of 24 gigahertz.

Townes later worked with Arthur L. Schawlow to describe the principle of the *optical maser*, or *laser*, which Theodore H. Maiman first demonstrated in 1960. For their research in this field Townes, Basov, and Prokhorov were awarded the Nobel Prize in Physics in 1964.

Technology

The maser is based on the principle of stimulated emission proposed by Albert Einstein in 1917. When atoms have been put into an excited energy state, they can amplify radiation at the proper frequency. By putting such an amplifying medium in a resonant cavity, feedback is created that can produce coherent radiation.

Some Common Types of Masers

- Atomic beam masers
- Ammonia maser
- Free Electron Maser
- Hydrogen maser
- Gas masers
- Rubidium maser
- Solid State masers
- Ruby maser

The dual noble gas maser is an example of a masing medium which is nonpolar.

Uses

Masers serve as high precision frequency references. These "atomic frequency standards" are one of the many forms of atomic clocks. They are also used as electronic amplifiers in radio telescopes. Masers are being developed as directed-energy weapons.

Hydrogen Maser

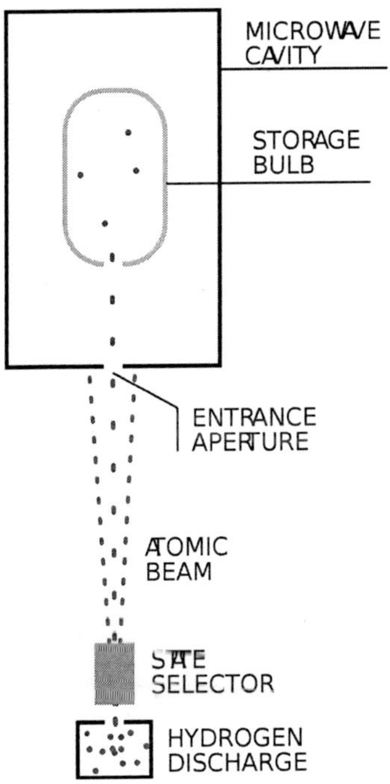

Figure: *A hydrogen maser.*

Today, the most important type of maser is the hydrogen maser which is currently used as an atomic frequency standard. Together with other types of atomic clocks, they constitute the "Temps Atomique International" or TAI. This is the international time scale, which is coordinated by the Bureau International des Poids et Mesures, or BIPM.

It was Norman Ramsey and his colleagues who first realised this device. Today's masers are identical to the original design. The maser oscillation relies on stimulated emission between two hyperfine levels of atomic hydrogen. Here is a brief description of how it works:

- First, a beam of atomic hydrogen is produced. This is done by submitting the gas at low pressure to an RF discharge.
- The next step is "state selection"—in order to get some stimulated emission, it is necessary to create a population inversion of the atoms. This is done in a way that is very similar to the famous Stern-Gerlach experiment. After passing through an aperture and a magnetic field, many of the atoms in the beam are left in the upper energy level of the lasing transition. From this state, the atoms can decay to the lower state and emit some microwave radiation.
- A high quality factor microwave cavity confines the microwaves and reinjects them repeatedly into the atom beam. The stimulated emission amplifies the microwaves on each pass through the beam. This combination of amplification and feedback is what defines all oscillators. The resonant frequency of the microwave cavity is exactly tuned to the hyperfine structure of hydrogen: 1420 405 751.768 Hz.
- A small fraction of the signal in the microwave cavity is coupled into a coaxial cable and then sent to a coherent receiver.
- The microwave signal coming out of the maser is very weak (a few pW). The frequency of the signal is fixed and *extremely* stable. The coherent receiver is used to amplify the signal and change the frequency. This is done using a series of phase-locked loops and a high performance quartz oscillator.

Astrophysical Masers

Maser-like stimulated emission also occurs in nature in interstellar space, and is frequently called superradiant emission to distinguish it from laboratory masers. Such emission is observed from molecules such as water (H_2O), hydroxyl radicals (OH), methanol (CH_3OH), formaldehyde (CH_2O), and silicon monoxide (SiO). Water molecules in star-forming regions can undergo a population inversion and emit radiation at 22 GHz, creating the brightest spectral line in the radio universe. Some water masers also emit radiation from a vibrational mode at 96 GHz.

Extremely powerful masers, associated with active galactic nuclei, are known as megamasers and are up to a million times more powerful than stellar masers.

Terminology

The meaning of the term *maser* has changed slightly since its introduction. Initially the acronym was universally given as "microwave amplification by stimulated emission of radiation," which described devices which emitted in the microwave region of the electromagnetic spectrum.

The principle and concept of stimulated emission has since been extended to more devices and frequencies. Thus the original acronym is sometimes modified, as suggested by Charles H. Townes, to "*molecular* amplification by stimulated emission of radiation."

Some have asserted that Townes's efforts to extend the acronym in this way were primarily motivated by the desire to increase the importance of his invention, and his reputation in the scientific community.

When the laser was developed, Townes and Schawlow and their colleagues at Bell Labs pushed the use of the term *optical maser*, but this was largely abandoned in favour of *laser*, coined by their rival Gordon Gould. In modern usage, devices that emit in the X-ray through infrared portions of the spectrum are typically called lasers, and devices that emit in the microwave region and below are commonly called *masers*, regardless of whether they emit microwaves or other frequencies.

Gould originally proposed distinct names for devices that emit in each portion of the spectrum, including *grasers* (gamma ray lasers), *xasers* (x-ray asers), *uvasers* (ultraviolet lasers), *lasers* (visiblelasers), *irasers* (infrared lasers), *masers* (microwave masers), and *rasers* (RF masers). Most of these terms never caught on, however, and all have now become (apart from in science fiction) obsolete except for *maser* and *laser*.

Laser

A laser is a device that emits light (electromagnetic radiation) through a process of optical amplification based on the stimulated emission of photons. The term "laser" originated as an acronym for *Light Amplification by Stimulated Emission of Radiation*. The emitted laser light is notable for its high degree of spatial and temporal coherence, unattainable using other technologies. Spatial coherence typically is expressed through the output being a narrow beam which is diffraction-limited, often a so-called "pencil beam."

Laser beams can be focused to very tiny spots, achieving a very high irradiance. Or they can be launched into a beam of very low divergence in order to concentrate their power at a large distance. Temporal (or longitudinal) coherence implies a polarized wave at a single frequency whose phase is correlated over a relatively large distance (the coherence length) along the beam. A beam produced by a thermal or other incoherent light source has an instantaneous amplitude and phase which vary randomly with respect to time and position, and thus a very short coherence length.Most so-called "single wavelength" lasers actually produce radiation in several *modes* having slightly different frequencies (wavelengths), often not in a single polarization. And although temporal coherence implies monochromaticity, there are even lasers that emit a broad spectrum of light, or emit different wavelengths of light simultaneously.

There are some lasers which are not single spatial mode and consequently their light beams diverge more than required by the diffraction limit. However all such devices are classified as "lasers" based on their method of producing that light: stimulated emission. Lasers are employed in applications where light of the required spatial or temporal coherence could not be produced using simpler technologies.

Terminology

The word *laser* started as an acronym for "light amplification by stimulated emission of radiation"; in modern usage "light" broadly denotes electromagnetic radiation of any frequency, not only visible light, hence infrared *laser*, *ultraviolet laser*, *X-ray laser*, and so on. Because the microwave predecessor of the laser, the maser, was developed first, devices of this sort operating at microwave and radio frequencies are referred to as "masers" rather than "microwave lasers" or "radio lasers". In the early technical literature, especially at Bell Telephone Laboratories, the laser was called an optical maser; this term is now obsolete.

A laser which produces light by itself is technically an optical oscillator rather than an optical amplifier as suggested by the acronym. It has been humorously noted that the acronym LOSER, for "light oscillation by stimulated emission of radiation," would have been more correct. With the widespread use of the original acronym as a common noun, actual optical amplifiers have come to be referred to as "laser amplifiers", notwithstanding the apparent redundancy in that designation.

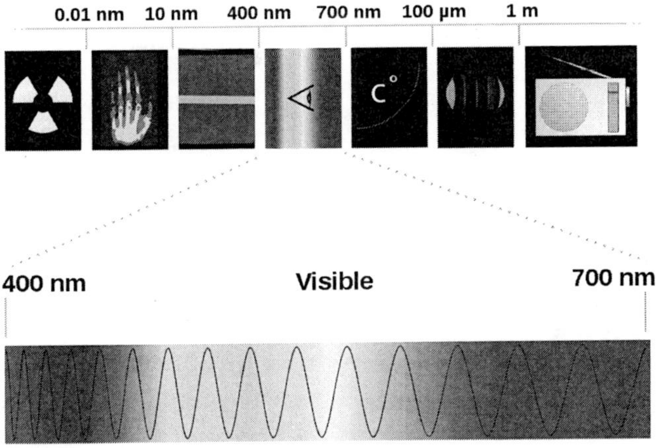

Figure: *From left to right: gamma rays, X-rays, ultraviolet rays, visible spectrum, infrared, microwaves, radio waves. Bottom: enlargement of visible spectrum from violet (400nm) to red (700nm).*

The back-formed verb *to lase* is frequently used in the field, meaning "to produce laser light," especially in reference to the gain medium of a laser; when a laser is operating it is said to be "lasing." Further use of the words *laser* and *maser* in an extended sense, not referring to laser technology or devices, can be seen in usages such as *astrophysical maser* and *atom laser*.

Design

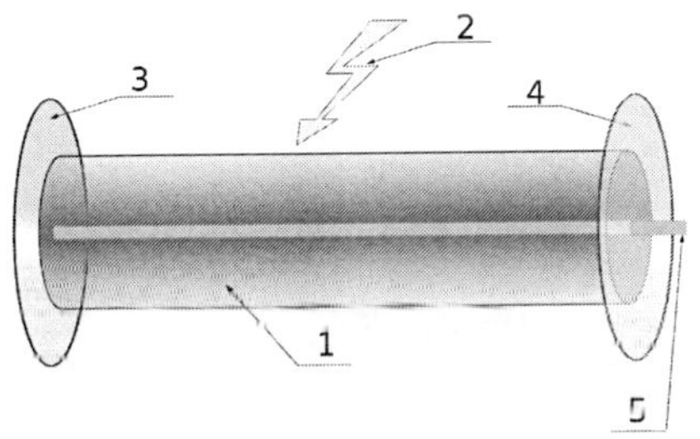

Figure: *Principal components: 1. Gain medium; 2. Laser pumping energy; 3. High reflector; 4. Output coupler; 5. Laser beam;*

A laser consists of a gain medium inside a highly reflective optical cavity, as well as a means to supply energy to the gain medium. The gain medium is a material with properties that allow it to amplify light by stimulated emission. In its simplest form, a cavity consists of two mirrors arranged such that light bounces back and forth, each time passing through the gain medium. Typically one of the two mirrors, the output coupler, is partially transparent. The output laser beam is emitted through this mirror.

Light of a specific wavelength that passes through the gain medium is amplified (increases in power); the surrounding mirrors ensure that most of the light makes many passes through the gain medium, being amplified repeatedly. Part of the light that is between the mirrors (that is, within the cavity) passes through the partially transparent mirror and escapes as a beam of light.

The process of supplying the energy required for the amplification is called pumping. The energy is typically supplied as an electrical current or as light at a different wavelength. Such light may be provided by a flash lamp or perhaps another laser. Most practical lasers contain additional elements that affect properties such as the wavelength of the emitted light and the shape of the beam.

Chapter 3

Fundamental Terms and Concepts

Matter

Matter is a general term for the substance of which all physical objects consist. Typically, matter includes atoms and other particles which have mass. A common way of defining *matter* is as anything that has mass and occupies volume. However, different fields use the term in different and sometimes incompatible ways; there is no single agreed scientific meaning of the word "matter".

For much of the history of the natural sciences people have contemplated the exact nature of matter. The idea that matter was built of discrete building blocks, the so-called *particulate theory of matter*, was first put forward by the Greek philosophers Leucippus (~490 BC) and Democritus (~470–380 BC). Over time an increasingly fine structure for matter was discovered: objects are made from molecules, molecules consist of atoms, which in turn consist of interacting subatomic particles like protons and electrons.

Matter is commonly said to exist in four *states* (or *phases*): solid, liquid, gas and plasma. However, advances in experimental techniques have realised other phases, previously only theoretical constructs, such as Bose–Einstein condensates and fermionic condensates. A focus on an elementary-particle view of matter also leads to new phases of matter, such as the quark–gluon plasma.

In physics and chemistry, matter exhibits both wave-like and particle-like properties, the so-called wave–particle duality.

In the realm of cosmology, extensions of the term *matter* are invoked to include dark matter and dark energy, concepts introduced to explain some odd phenomena of the observable universe, such as the galactic rotation curve. These exotic forms of "matter" do not refer to matter as "building blocks", but rather to currently poorly understood forms of mass and energy.

Figure: *Matter is usually classified into three classical states, with plasma sometimes added as a fourth state. From top to bottom: quartz (solid), water (liquid), nitrogen dioxide (gas), and a plasma globe (plasma).*

Historical Development

Origins

The pre-Socratics were among the first recorded speculators about the underlying nature of the visible world. Thales (c. 624 BC–c. 546 BC) regarded water as the fundamental material of the world. Anaximander (c. 610 BC–c. 546 BC) posited that the basic material was wholly characterless or limitless: the Infinite (*apeiron*). Anaximenes (flourished 585 BC, d. 528 BC) posited that the basic stuff was *pneuma* or air. Heraclitus (c. 535–c. 475 BC) seems to say the basic element is fire, though perhaps he means that all is change. Empedocles (c. 490–430 BC) spoke of four elements of which everything was made: earth, water, air, and fire. Meanwhile, Parmenides argued that change does not exist, and Democritus argued that everything is composed of minuscule, inert bodies of all shapes called atoms, a philosophy called atomism. All of these notions had deep philosophical problems.

Aristotle (384 BC – 322 BC) was the first to put the conception on a sound philosophical basis, which he did in his natural philosophy, especially in *Physics* book I. He adopted as reasonable suppositions the four Empedoclean elements, but added a fifth, aether. Nevertheless these elements are not basic in Aristotle's mind. Rather they, like everything else in the visible world, are composed of the basic *principles* matter and form.

The word Aristotle uses for matter, Qλη (*hyle* or *hule*), can be literally translated as wood or timber, that is, "raw material" for building. Indeed, Aristotle's conception of matter is intrinsically linked to something being made or composed. In other words, in contrast to the early modern conception of matter as simply occupying space, matter for Aristotle is definitionally linked to process or change: matter is what underlies a change of substance.

For example, a horse eats grass: the horse changes the grass into itself; the grass as such does not persist in the horse, but some aspect of it—its matter—does. The matter is not specifically described (e.g., as atoms), but consists of whatever persists in the change of substance from grass to horse. Matter in this understanding does not exist independently (i.e., as a substance), but exists interdependently (i.e., as a "principle") with form and only insofar as it underlies change. It can be helpful to conceive of the relationship of matter and form as very similar to that between parts and whole. For Aristotle, matter as such can only *receive* actuality from form; it has no activity or actuality in itself, similar to the way that parts as such only have their existence *in* a whole (otherwise they would be independent wholes).

Early Modernity

René Descartes (1596–1650) originated the modern conception of matter. He was primarily a geometre. Instead of, like Aristotle, deducing the existence of matter from the physical reality of change, Descartes arbitrarily postulated matter to be an abstract, mathematical substance that occupies space:

> *So, extension in length, breadth, and depth, constitutes the nature of bodily substance; and thought constitutes*

the nature of thinking substance. And everything else which can be attributed to body presupposes extension, and is only a mode of that which is extended
— René Descartes, Principles of Philosophy

For Descartes, matter has only the property of extension, so its only activity aside from locomotion is to exclude other bodies: this is the mechanical philosophy. Descartes makes an absolute distinction between mind, which he defines as unextended, thinking substance, and matter, which he defines as unthinking, extended substance. They are independent things. In contrast, Aristotle defines matter and the formal/forming principle as complementary *principles* which together compose one independent thing (substance). In short, Aristotle defines matter (roughly speaking) as what things are actually made of (with a *potential* independent existence), but Descartes elevates matter to an actual independent thing in itself.

The continuity and difference between Descartes' and Aristotle's conceptions is noteworthy. In both conceptions, matter is passive or inert. In the respective conceptions matter has different relationships to intelligence. For Aristotle, matter and intelligence (form) exist together in an interdependent relationship, whereas for Descartes, matter and intelligence (mind) are definitionally opposed, independent substances.

Descartes' justification for restricting the inherent qualities of matter to extension is its permanence, but his real criterion is not permanence (which equally applied to colour and resistance), but his desire to use geometry to explain all material properties. Like Descartes, Hobbes, Boyle, and Locke argued that the inherent propeties of bodies were limited to extension, and that so-called secondary qualities, like colour, were only products of human perception.

Isaac Newton (1643–1727) inherited Descartes' mechanical conception of matter. In the third of his "Rules of Reasoning in Philosophy," Newton lists the universal qualities of matter as "extension, hardness, impenetrability, mobility, and inertia." Similarly in *Optics* he conjectures that God created matter as "solid, massy, hard, impenetrable, movable particles", which were "even so very hard as never to wear or break in pieces." The "primary" properties of matter were amenable to mathematical description, unlike "secondary" qualities such as colour or taste. Like Descartes, Newton rejected the essential nature of secondary qualities. Newton developed Descartes' notion of matter by restoring to matter intrinsic properties in addition to extension (at least on a limited basis), such as mass. Newton's use of gravitational force, which worked "at a distance," effectively repudiated Descartes' mechanics, in which interactions happened exclusively by contact. Though Newton's gravity would seem to be a *power* of bodies, Newton himself did not admit it to be an essential property of matter. Carrying the logic forward more consistently, Joseph Priestley argued that corporeal properties transcend contact mechanics: chemical properties require the *capacity* for attraction. He argued matter has other inherent powers besides the so-called primary qualities of Descartes, et al.

Since Priestley's time, there has been a massive expansion in knowledge of the constituents of the material world (viz., molecules, atoms, subatomic particles), but there has been no further development in the *definition* of matter. Rather the question has been set aside. Noam Chomskysummarises the situation that has prevailed since that time:

> *What is the concept of body that finally emerged?[...] The answer is that there is no clear and definite conception of body.[...] Rather, the material world is whatever we discover it to be, with whatever properties it must be assumed to have for the purposes of explanatory theory. Any intelligible theory that offers genuine explanations and that can be assimilated to the core notions of physics becomes part of the theory of the material world, part of our account of body. If we have such a theory in some domain, we seek to assimilate it to the core notions of physics, perhaps modifying these notions as we carry out this enterprise.*
>
> — *Noam Chomsky, 'Language and problems of knowledge: the Managua lectures, p. 144*

So matter is whatever physics studies and the object of study of physics is matter: there is no independent general definition of matter, apart from its fitting into the methodology of measurement and controlled experimentation. In sum, the boundaries between what constitutes matter and everything else remains as vague as the demarcation problem of delimiting science from everything else.

Late Nineteenth and Early Twentieth Century

In the 19th century, following the development of the periodic table, and of atomic theory, atoms were seen as being the fundamental constituents of matter; atoms formed molecules and compounds.

The common definition in terms of occupying space and having mass is in contrast with most physical and chemical definitions of matter, which rely instead upon its structure and upon attributes not necessarily related to volume and mass. At the turn of the nineteenth century, the knowledge of matter began a rapid evolution.

Aspects of the Newtonian view still held sway. James Clerk Maxwell discussed matter in his work *Matter and Motion*. He carefully separates "matter" from space and time, and defines it in terms of the object referred to in Newton's first law of motion.

However, the Newtonian picture was not the whole story. In the 19th century, the term "matter" was actively discussed by a host of scientists and philosophers, and a brief outline can be found in Levere. A textbook discussion from 1870 suggests matter is what is made up of atoms:

> *Three divisions of matter are recognised in science: masses, molecules and atoms.*
>
> *A Mass of matter is any portion of matter appreciable by the senses.*
>
> *A Molecule is the smallest particle of matter into which a body can be divided without losing its identity.*
>
> *An Atom is a still smaller particle produced by division of a molecule.*

Rather than simply having the attributes of mass and occupying space, matter was held to have chemical and electrical properties. The famous physicist J. J. Thomson wrote about the "constitution of matter" and was concerned with the possible connection between matter and electrical charge.

Later Developments

There is an entire literature concerning the "structure of matter", ranging from the "electrical structure" in the early 20th century, to the more recent "quark structure of matter", introduced today with the remark: *Understanding the quark structure of matter has been one of the most important advances in contemporary physics.* In this connection, physicists speak of *matter fields*, and speak of particles as "quantum excitations of a mode of the matter field". And here is a quote from de Sabbata and Gasperini: "With the word "matter" we denote, in this context, the sources of the interactions, that is spinor fields (like quarks and leptons), which are believed to be the fundamental components of matter, or scalar fields, like the Higgs particles, which are used to introduced mass in a gauge theory (and which, however, could be composed of more fundamental fermion fields)."

The modern conception of matter has been refined many times in history, in light of the improvement in knowledge of just *what* the basic building blocks are, and in how they interact.

In the late 19th century with the discovery of the electron, and in the early 20th century, with the discovery of the atomic nucleus, and the birth of particle physics, matter was seen as made up of electrons, protons and neutrons interacting to form atoms. Today, we know that even protons and neutrons are not indivisible, they can be divided into quarks, while electrons are part of a particle family called leptons. Both quarks and leptons are elementary particles, and are currently seen as being the fundamental constituents of matter.

These quarks and leptons interact through four fundamental forces: gravity, electromagnetism, weak interactions, and strong interactions. The Standard Model of particle physics is currently the best explanation for all of physics, but despite decades of efforts, gravity cannot yet be accounted for at the quantum-level; it is only described by classical physics (quantum gravity and graviton). Interactions between quarks and leptons are the result of an exchange of force-carrying particles (such as photons) between quarks and leptons. The force-carrying particles are not themselves building blocks. As one consequence, mass and energy (which cannot be created or destroyed) cannot always be related to matter (which can be created out of non-matter particles such as photons, or even out of pure energy, such as kinetic energy). Force carriers are usually not considered matter: the carriers of the electric force (photons) possess energy (Planck relation) and the carriers of the weak force (W and Z bosons) are massive, but neither are considered matter either. However, while these particles are not considered matter, they do contribute to the total mass of atoms, subatomic particles, and all systems which contain them.

The term "matter" is used throughout physics in a bewildering variety of contexts: for example, one refers to "condensed matter physics", "elementary matter", "partonic" matter, "dark" matter, "anti"-matter, "strange" matter, and "nuclear" matter. In discussions of matter and antimatter, normal matter has been referred to by Alfvén as *koinomatter*. It is fair to say that in physics, there is no broad consensus as to a general definition of matter, and the term "matter" usually is used in conjunction with a specifying modifier.

Definitions

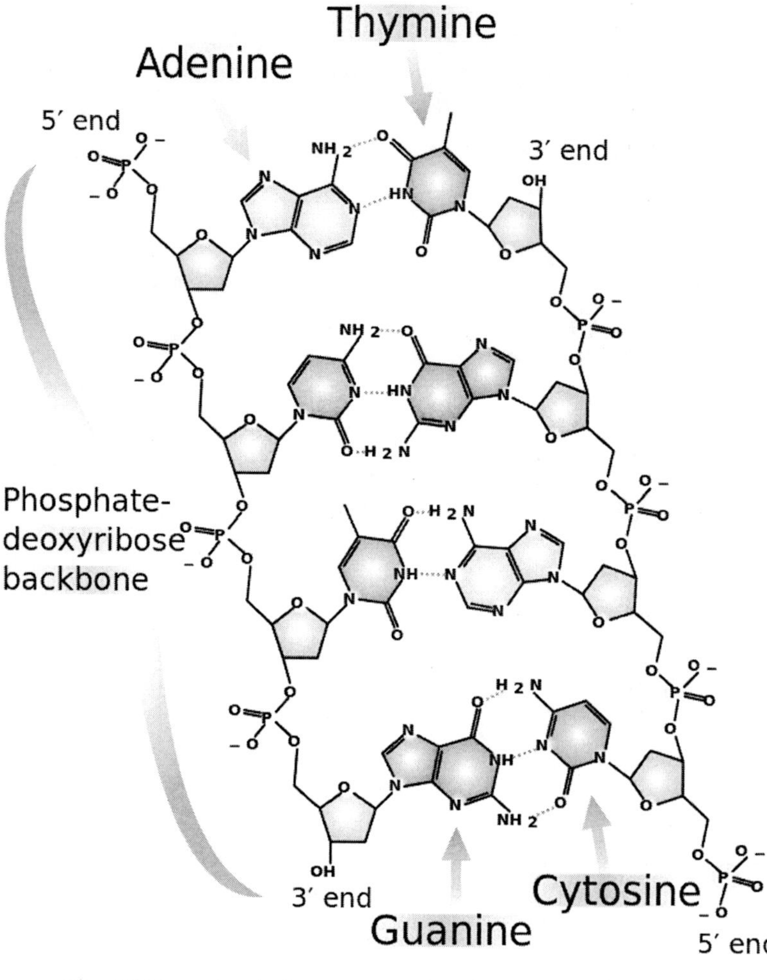

Figure: *The DNA molecule is an example of* matter *under the "atoms and molecules" definition.*

The common definition of matter is *anything that has both mass and volume (occupies space)*. For example, a car would be said to be made of matter, as it occupies space, and has mass.

The observation that matter occupies space goes back to antiquity. However, an explanation for why matter occupies space is recent, and is argued to be a result of the Pauli exclusion principle. Two particular examples where the exclusion principle clearly relates matter to the occupation of space are white dwarf stars and neutron stars, discussed further below.

Relativity

In the context of relativity, mass is not an additive quantity. Thus, in relativity usually a more general view is taken that it is not mass, but the energy–momentum tensor that quantifies the amount of matter. Matter therefore is anything that contributes to the energy–momentum of a system, that is, anything that is not purely gravity. This view is commonly held in fields that deal with general relativity such as cosmology.

Atoms and Molecules Definition

A definition of "matter" that is based upon its physical and chemical structure is: *matter is made up of atoms and molecules.* As an example, deoxyribonucleic acid molecules (DNA) are matter under this definition because they are made of atoms. This definition can be extended to include charged atoms and molecules, so as to include plasmas (gases of ions) and electrolytes (ionic solutions), which are not obviously included in the atoms and molecules definition. Alternatively, one can adopt the *protons, neutrons and electrons* definition.

Protons, Neutrons and Electrons Definition

A definition of "matter" more fine-scale than the atoms and molecules definition is: *matter is made up of what atoms and molecules are made of*, meaning anything made of positively charged protons, neutral neutrons, and negatively charged electrons. This definition goes beyond atoms and molecules, however, to include substances made from these building blocks that are *not* simply atoms or molecules, for example white dwarf matter — typically, carbon and oxygen nuclei in a sea of degenerate electrons. At a microscopic level, the constituent "particles" of matter such as protons, neutrons and electrons obey the laws of quantum mechanics and exhibit wave–particle duality. At an even deeper level, protons and neutrons are made up of quarks and the force fields (gluons) that bind them together.

Quarks and Leptons Definition

As may be seen from the above discussion, many early definitions of what can be called *ordinary matter* were based upon its structure or "building blocks". On the scale of elementary particles, a definition that follows this tradition can be stated as: *ordinary matter is everything that is composed of elementary fermions, namely quarks and leptons.* The connection between these formulations follows.

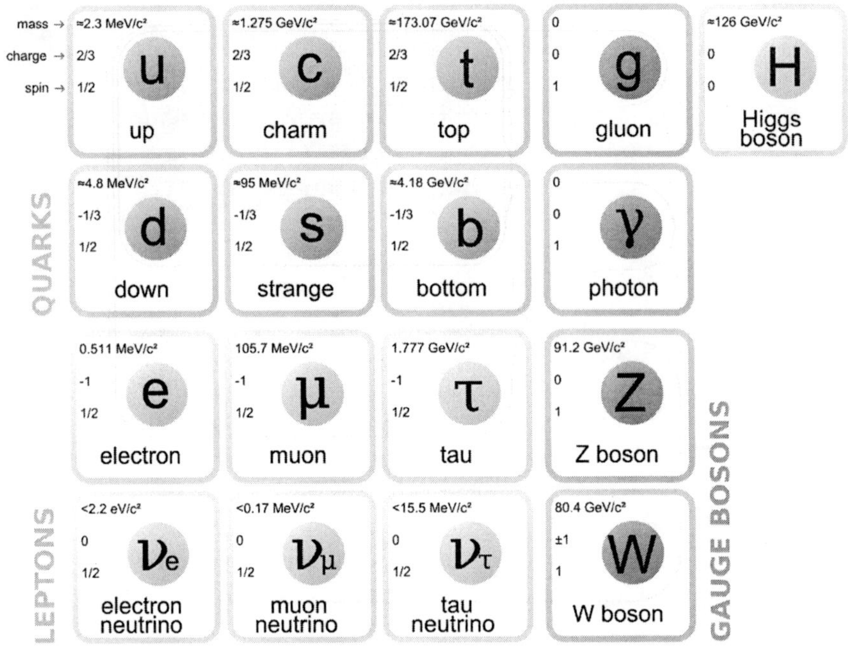

***Figure:** Under the "quarks and leptons" definition, the elementary and composite particles made of the quarks (in purple) and leptons (in green) would be "matter"; while the gauge bosons (in red) would not be "matter". However, interaction energy inherent to composite particles (for example, gluons involved in neutrons and protons) contribute to the mass of ordinary matter.*

Leptons (the most famous being the electron), and quarks (of which baryons, such as protons and neutrons, are made) combine to form atoms, which in turn form molecules. Because atoms and molecules are said to be matter, it is natural to phrase the definition as: ordinary *matter is anything that is made of the same things that atoms and molecules are made of.* (However, notice that one also can make from these building blocks matter that is *not* atoms or molecules.) Then, because electrons are leptons, and protons and neutrons are made of quarks, this definition in turn leads to the definition of matter as being "quarks and leptons", which are the two types of elementary fermions.

Carithers and Grannis state: *Ordinary matter is composed entirely of first-generation particles, namely the [up] and [down] quarks, plus the electron and its neutrino.* (Higher generations particles quickly decay into first-generation particles, and thus are not commonly encountered.)

This definition of ordinary matter is more subtle than it first appears. All the particles that make up ordinary matter (leptons and quarks) are elementary fermions, while all the force carriers are elementary bosons. The W and Z bosons that mediate the weak force are not made of quarks or leptons, and so are not ordinary matter, even if they have mass. In other words, mass is not something that is exclusive to ordinary matter.

The quark–lepton definition of ordinary matter, however, identifies not only the elementary building blocks of matter, but also includes composites made from the constituents (atoms and molecules, for example). Such composites contain an interaction energy that holds the constituents together, and may constitute the bulk of the mass of the composite. As an example, to a great extent, the mass of an atom is simply the sum of the masses of its constituent protons, neutrons and electrons. However, digging deeper, the protons and neutrons are made up of quarks bound together by gluon fields and these gluons fields contribute significantly to the mass of hadrons. In other words, most of what composes the "mass" of ordinary matter is due to the binding energy of quarks within protons and neutrons. For example, the sum of the mass of the three quarks in a nucleon is approximately 12.5 MeV/c^2, which is low compared to the mass of a nucleon (approximately 938 MeV/c^2). The bottom line is that most of the mass of everyday objects comes from the interaction energy of its elementary components.

Smaller Building Blocks?

The Standard Model groups matter particles into three generations, where each generation consists of two quarks and two leptons. The first generation is the *up* and *down* quarks, the *electron* and the electron *neutrino*; the second includes the *charm* and *strange* quarks, the *muon* and the *muon neutrino*; the third generation consists of the *top* and *bottom* quarks and the *tau* and *tau neutrino.*

The most natural explanation for this would be that quarks and leptons of higher generations are excited states of the first generations. If this turns out to be the case, it would imply that quarks and leptons are composite particles, rather than elementary particles.

Structure

In particle physics, fermions are particles which obey Fermi Dirac statistics. Fermions can be elementary, like the electron, or composite, like the proton and the neutron. In the Standard Model there are two types of elementary fermions: quarks and leptons, which are discussed next.

Quarks

Quarks are a particles of spin-$^{1}D_{2}$, implying that they are fermions. They carry an electric charge of “$^{1}D_{3}$ e (down-type quarks) or +$^{2}D_{3}$ e (up-type quarks). For comparison, an electron has a charge of “1 e. They also carry colour charge, which is the equivalent of the electric charge for the strong interaction. Quarks also undergo radioactive decay, meaning that they are subject to the weak interaction. Quarks are massive particles, and therefore are also subject to gravity.

Quark Properties

name	*symbol*	*spin*	*electric charge (e)*	*mass (MeV/c²)*	*mass comparable to*	*antiparticle*	*antiparticle symbol*
up-type quarks							
up	u	½	+⅔	1.5 to 3.3	~ 5 electrons	antiup	U
charm	c	½	+⅔	1160 to 1340	~ 1 proton	anticharm	C
top	t	½	+⅔	169,100 to 173,300	~ 180 protons or ~ 1 tungsten atom	antitop	T
down-type quarks							
down	d	½	−⅓	3.5 to 6.0	~ 10 electrons	antidown	D
strange	s	½	−⅓	70 to 130	~ 200 electrons	antistrange	S
bottom	b	½	−⅓	4130 to 4370	~ 5 protons	antibottom	B

Baryonic Matter

Figure: *Quark structure of a proton: 2 up quarks and 1 down quark.*

Baryons are strongly interacting fermions, and so are subject to Fermi-Dirac statistics. Amongst the baryons are the protons and neutrons, which occur in atomic nuclei, but many other unstable baryons exist as well. The termbaryon is usually used to refer to triquarks — particles made of three quarks. "Exotic" baryons made of four quarks and one antiquark are known as the pentaquarks, but their existence is not generally accepted.

Baryonic matter is the part of the universe that is made of baryons (including all atoms). This part of the universe does not include dark energy, dark matter, black holes or various forms of degenerate matter, such as compose white dwarf stars and neutron stars. Microwave light seen by Wilkinson Microwave Anisotropy Probe (WMAP), suggests that only about 4.6% of that part of the universe within range of the best telescopes (that is, matter that may be visible because light could reach us from it), is made of baryionic matter. About 23% is dark matter, and about 72% is dark energy.

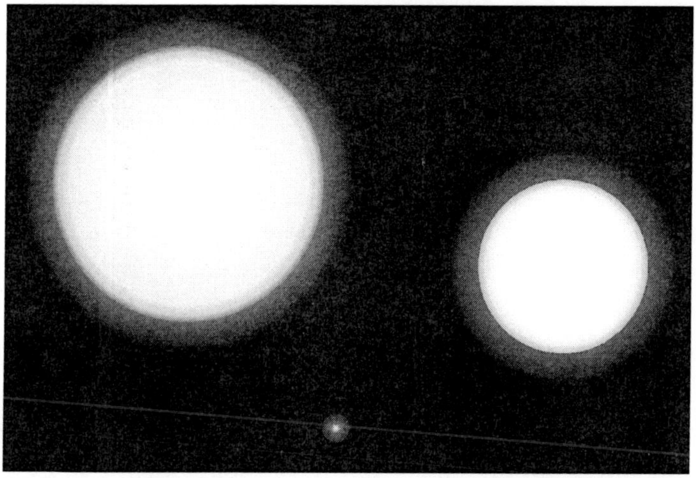

Figure: *A comparison between the white dwarf IK Pegasi B (centre), its A-class companion IK Pegasi A (left) and the Sun (right). This white dwarf has a surface temperature of 35,500 K.*

Degenerate Matter

In physics, degenerate matter refers to the ground state of a gas of fermions at a temperature near absolute zero. The Pauli exclusion principle requires that only two fermions can occupy a quantum state, one spin-up and the other spin-down. Hence, at zero temperature, the fermions fill up sufficient levels to accommodate all the available fermions, and for the case of many fermions the maximum kinetic energy called the Fermi energy and the pressure of the gas becomes very large and dependent upon the number of fermions rather than the temperature, unlike normal states of matter. Degenerate matter is thought to occur during the evolution of heavy stars. The demonstration by Subrahmanyan Chandrasekhar that white dwarf stars have a maximum allowed mass because of the exclusion principle caused a revolution in the theory of star evolution. Degenerate matter includes the part of the universe that is made up of neutron stars and white dwarfs.

Strange Matter

Strange matter is a particular form of quark matter, usually thought of as a 'liquid' of up, down, and strange quarks. It is to be contrasted with nuclear matter, which is a liquid of neutrons and protons (which themselves are built out of up and down quarks), and with non-strange quark matter, which is a quark liquid containing only up and down quarks. At high enough density, strange matter is expected to be colour superconducting. Strange matter is hypothesized to occur in the core of neutron stars, or, more speculatively, as isolated droplets that may vary in size from femtometres (strangelets) to kilometres (quark stars).

Two Meanings of the Term "Strange Matter"

In particle physics and astrophysics, the term is used in two ways, one broader and the other more specific.

1. The broader meaning is just quark matter that contains three flavours of quarks: up, down, and strange. In this definition, there is a critical pressure and an associated critical density, and when nuclear matter (made of protons and neutrons) is compressed beyond this density, the protons and neutrons dissociate into quarks, yielding quark matter (probably strange matter).
2. The narrower meaning is quark matter that is *more stable than nuclear matter*. The idea that this could happen is the "strange matter hypothesis" of Bodmer and Witten. In this definition, the critical pressure is zero: the true ground state of matter is *always* quark matter. The nuclei that we see in the matter around us, which are droplets of nuclear matter, are actually metastable, and given enough time (or the right external stimulus) would decay into droplets of strange matter, i.e. strangelets.

Leptons

Leptons are particles of spin-^{1}D $_{2}$, meaning that they are fermions. They carry an electric charge of "1 e(charged leptons) or 0 e (neutrinos). Unlike quarks, leptons do not carry colour charge, meaning that they do not experience the strong interaction. Leptons also undergo radioactive decay, meaning that they are subject to the weak interaction. Leptons are massive particles, therefore are subject to gravity.

Lepton properties							
name	*symbol*	*spin*	*electric charge (e)*	*mass (MeV/c²)*	*mass comparable to*	*antiparticle*	*antiparticle symbol*
charged leptons[71]							
electron	e−	½	−1	0.5110	1 electron	antielectron	e+
muon	μ−	½	−1	105.7	~ 200 electrons	antimuon	μ+
tau	τ−	½	−1	1,777	~ 2 protons	antitau	τ+
neutrinos[72]							
electron neutrino	ν_e	½	0	< 0.000460	< 1⁄1000 electron	electron antineutrino	ν_e
muon neutrino	ν_μ	½	0	< 0.19	< ½ electron	muon antineutrino	ν_μ
tau neutrino	ν_τ	½	0	< 18.2	< 40 electrons	tau antineutrino	ν_τ

Phases

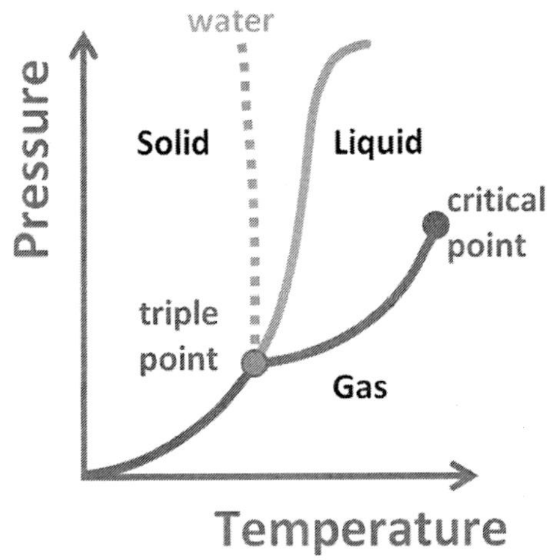

Figure: *Phase diagram for a typical substance at a fixed volume. Vertical axis is* Pressure, *horizontal axis is* Temperature. *The green line marks the freezing point (above the green line is* solid, *below it is* liquid*) and the blue line the boiling point (above it is* liquid *and below it is* gas*). So, for example, at higher* T, *a higher* P *is necessary to maintain the substance in liquid phase. At the triple point the three phases; liquid, gas and solid; can coexist. Above the critical point there is no detectable difference between the phases. The dotted line shows the anomalous behaviour of water: ice melts at constant temperature with increasing pressure.*

In bulk, matter can exist in several different forms, or states of aggregation, known as *phases*, depending on ambient pressure, temperature and volume. A phase is a form of matter that has a relatively uniform chemical composition and physical properties (such as density, specific heat, refractive index, and so forth). These phases include the three familiar ones (solids, liquids, and gases), as well as more exotic states of matter (such as plasmas, superfluids, supersolids, Bose–Einstein condensates, ...).

A *fluid* may be a liquid, gas or plasma. There are also paramagnetic and ferromagnetic phases of magnetic materials. As conditions change, matter may change from one phase into another. These phenomena are called phase transitions, and are studied in the field of thermodynamics.

In nanomaterials, the vastly increased ratio of surface area to volume results in matter that can exhibit properties entirely different from those of bulk material, and not well described by any bulk phase. Phases are sometimes called *states of matter*, but this term can lead to confusion with thermodynamic states. For example, two gases maintained at different pressures are in different *thermodynamic states* (different pressures), but in the same *phase* (both are gases).

Antimatter

In particle physics and quantum chemistry, antimatter is matter that is composed of the antiparticles of those that constitute ordinary matter. If a particle and its antiparticle come into contact with each other, the two annihilate; that is, they may both be converted into other particles with equal energy in accordance with Einstein's equation $E = mc^2$.

These new particles may be high-energy photons (gamma rays) or other particle–antiparticle pairs. The resulting particles are endowed with an amount of kinetic energy equal to the difference between the rest mass of the products of the annihilation and the rest mass of the original particle-antiparticle pair, which is often quite large. Antimatter is not found naturally on Earth, except very briefly and in vanishingly small quantities (as the result of radioactive decay or cosmic rays). This is because antimatter which came to exist on Earth outside the confines of a suitable physics laboratory would almost instantly meet the ordinary matter that Earth is made of, and be annihilated.

Antiparticles and some stable antimatter (such as antihydrogen) can be made in tiny amounts, but not in enough quantity to do more than test a few of its theoretical properties.

There is considerable speculation both in science and science fiction as to why the observable universe is apparently almost entirely matter, and whether other places are almost entirely antimatter instead. In the early universe, it is thought that matter and antimatter were equally represented, and the disappearance of antimatter requires an asymmetry in physical laws called the charge parity (orCP symmetry) violation.

CP symmetry violation can be obtained from the Standard Model, but at this time the apparent asymmetry of matter and antimatter in the visible universe is one of the great unsolved problems in physics. Possible processes by which it came about are explored in more detail under baryogenesis.

Other Types of Matter

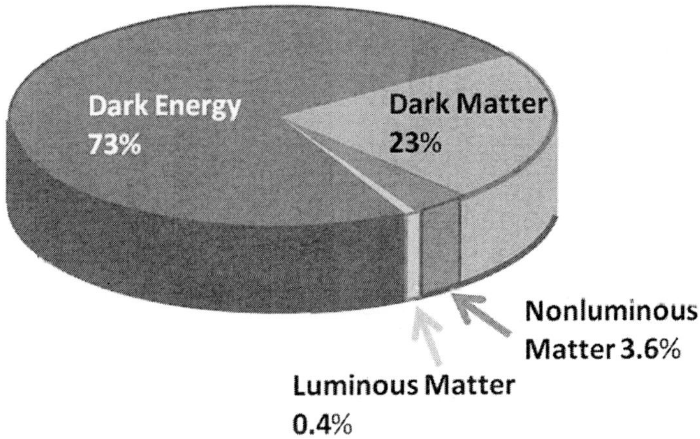

Figure: *Pie chart showing the fractions of energy in the universe contributed by different sources.* Ordinary matter *is divided into* luminous matter *(the stars and luminous gases and 0.005% radiation) and* nonluminous matter *(intergalactic gas and about 0.1% neutrinos and 0.04% supermassive black holes). Ordinary matter is uncommon. Modelled after Ostriker and Steinhardt.*

Ordinary matter, in the quarks and leptons definition, constitutes about 4% of the energy of the observable universe. The remaining energy is theorized to be due to exotic forms, of which 23% is dark matter and 73% is dark energy.

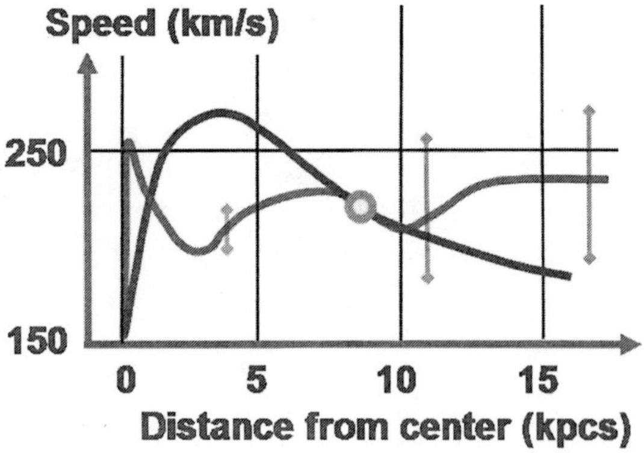

Figure: *Galaxy rotation curve for the Milky Way. Vertical axis is speed of rotation about the galactic centre. Horizontal axis is distance from the galactic centre. The sun is marked with a yellow ball. The observed curve of speed of rotation is blue. The predicted curve based upon stellar mass and gas in the Milky Way is red. The difference is due to dark matter or perhaps a modification of the law of gravity. Scatter in observations is indicated roughly by gray bars.*

Dark Matter

In astrophysics and cosmology, dark matteris matter of unknown composition that does not emit or reflect enough electromagnetic radiation to be observed directly, but whose presence can be inferred from gravitational effects on visible matter. Observational evidence of the early universe and the big bang theory require that this matter have energy and mass, but is not composed of either elementary fermions (as above) OR gauge bosons. The commonly accepted view is that most of the dark-matter is non-baryonic in nature. As such, it is composed of particles as yet unobserved in the laboratory. Perhaps they are supersymmetric particles, which are not Standard Model particles, but relics formed at very high energies in the early phase of the universe and still floating about.

Dark Energy

In cosmology, dark energy is the name given to the antigravitating influence that is accelerating the rate of expansion of the universe. It is known not to be composed of known particles like protons, neutrons or electrons, nor of the particles of dark matter, because these all gravitate. Fully 70% of the matter density in the universe appears to be in the form of dark energy. Twenty-six percent is dark matter. Only 4% is ordinary matter. So less than 1 part in 20 is made out of matter we have observed experimentally or described in the standard model of particle physics. Of the other 96%, apart from the properties just mentioned, we know absolutely nothing.

Exotic Matter

Exotic matter is a hypothetical concept of particle physics. It covers any material which violates one or more classical conditions or is not made of known baryonic particles. Such materials would possess qualities like negative mass or being repelled rather than attracted by gravity.

Energy

In physics, energy (Ancient Greek:vÝñãåéá *energeia* "activity, operation") is an indirectly observed quantity. It is often understood as the ability a physical system has to do work on other physical systems. Since work is defined as a force acting through a distance (a length of space), energy is always equivalent to the ability to exert pulls or pushes against the basic forces of nature, along a path of a certain length.

Figure: *Energy transformation; In a typical lightning strike, 500 megajoules of electric potential energy is converted into the same amount of energy in other forms, most notably light energy, sound energy and thermal energy.*

The total energy contained in an object is identified with its mass, and energy (like mass), cannot be created or destroyed. When matter (ordinary material particles) is changed into energy (such as energy of motion, or into radiation), the mass of the system does not change through the transformation process. However, there may be mechanistic limits as to how much of the matter in an object may be changed into other types of energy and thus into work, on other systems. Energy, like mass, is a scalar physical quantity. In the International System of Units (SI), energy is measured in joules, but in many fields other units, such as kilowatt-hours and kilocalories, are customary. All of these units translate to units of work, which is always defined in terms of forces and the distances that the forces act through.

A system can transfer energy to another system by simply transferring matter to it (since matter is equivalent to energy, in accordance with its mass). However, when energy is transferred by means other than matter-transfer, the transfer produces changes in the second system, as a result of work done on it. This work manifests itself as the effect of force(s) applied through distances within the target system. For example, a system can emit energy to another by transferring (radiating) electromagnetic energy, but this creates forces upon the particles that absorb the radiation. Similarly, a system may transfer energy to another by physically impacting it, but that case the energy of motion in an object, called kinetic energy, results in forces acting over distances (new energy) to appear in another object that is struck. Transfer of thermal energy by heat occurs by both of these mechanisms: heat can be transferred by electromagnetic radiation, or by physical contact in which direct particle-particle impacts transfer kinetic energy.

Energy may be stored in systems without being present as matter, or as kinetic or electromagnetic energy. Stored energy is created whenever a particle has been moved through a field it interacts with (requiring a force to do so), but the energy to accomplish this is stored as a new position of the particles in the field—a configuration that must be "held" or fixed by a different type of force (otherwise, the new configuration would resolve itself by the field pushing or pulling the particle back toward its previous position). This type of energy "stored" by force-fields and particles that have been forced into a new physical configuration in the field by doing work on them by another system, is referred to as potential energy. A simple example of potential energy is the work needed to lift an object in a gravity field, up to a support. Each of the basic forces of nature is associated with a different type of potential energy, and all types of potential energy (like all other types of energy) appears as system mass, whenever present. For example, a compressed spring will be slightly more massive than before it was compressed. Likewise, whenever energy is transferred between systems by any mechanism, an associated mass is transferred with it.

Any form of energy may be transformed into another form. For example, all types of potential energy are converted into kinetic energy when the objects are given freedom to move to different position (as for example, when an object falls off a support). When energy is in a form other than thermal energy, it may be transformed with good or even perfect efficiency, to any other type of energy, including electricity or production of new particles of matter. With thermal energy, however, there are often limits to the efficiency of the conversion to other forms of energy, as described by the second law of thermodynamics.

In all such energy transformation processes, the total energy remains the same, and a transfer of energy from one system to another, results in a loss to compensate for any gain. This principle, the conservation of energy, was first postulated in the early 19th century, and applies to any isolated system. According to Noether's theorem, the conservation of energy is a consequence of the fact that the laws of physics do not change over time.

Although the total energy of a system does not change with time, its value may depend on the frame of reference. For example, a seated passenger in a moving airplane has zero kinetic energy relative to the airplane, but non-zero kinetic energy (and higher total energy) relative to the Earth.

History

The word *energy* derives from the Greek *energeia*, which possibly appears for the first time in the work of Aristotle in the 4th century BCE.

The concept of energy emerged out of the idea of *vis viva* (living force), which Gottfried Leibniz defined as the product of the mass of an object and its velocity squared; he believed that total *vis viva* was conserved. To account for slowing due to friction, Leibniz theorized that thermal energy consisted of the random motion of the constituent parts of matter, a view shared by Isaac Newton, although it would be more than a century until this was generally accepted. In 1807, Thomas Young was possibly the first to use the term "energy" instead of *vis viva*, in its modern sense. Gustave-Gaspard Coriolis described "kinetic energy" in 1829 in its modern sense, and in 1853, William Rankine coined the term "potential energy". It was argued for some years whether energy was a substance (the caloric) or merely a physical quantity, such as momentum.

William Thomson (Lord Kelvin) amalgamated all of these laws into the laws of thermodynamics, which aided in the rapid development of explanations of chemical processes by Rudolf Clausius, Josiah Willard Gibbs, and Walther Nernst. It also led to a mathematical formulation of the concept of entropyby Clausius and to the introduction of laws of radiant energy by Jožef Stefan.

During a 1961 lecture for undergraduate students at the California Institute of Technology, Richard Feynman, a celebrated physics teacher and Nobel Laureate, said this about the concept of energy:

> There is a fact, or if you wish, a *law*, governing all natural phenomena that are known to date. There is no known exception to this law—it is exact so far as we know. The law is called the *conservation of energy*. It states that there is a certain quantity, which we call energy, that does not change in manifold changes which nature undergoes. That is a most abstract idea, because it is a mathematical principle; it says that there is a numerical quantity which does not change when something happens. It is not a description of a mechanism, or anything concrete; it is just a strange fact that we can calculate some number and when we finish watching nature go through her tricks and calculate the number again, it is the same.
>
> —*The Feynman Lectures on Physics*

Since 1918 it has been known that the law of conservation of energy is the direct mathematical consequence of the translational symmetry of the quantity conjugate to energy, namely time. That is, energy is conserved because the laws of physics do not distinguish between different instants of time.

Energy in Various Contexts

The concept of energy and its transformations is useful in explaining and predicting most natural phenomena. The *direction* of transformations in energy (what kind of energy is transformed to what other kind) is often described by entropy (equal energy spread among all available degrees of freedom) considerations, as in practice all energy transformations are permitted on a small scale, but certain larger transformations are not permitted because it is statistically unlikely that energy or matter will randomly move into more concentrated forms or smaller spaces.

The concept of energy is widespread in all sciences.

- In the context of chemistry, energy is an attribute of a substance as a consequence of its atomic, molecular or aggregate structure. Since a chemical transformation is accompanied by a change in one or more of these kinds of structure, it is invariably accompanied by an increase or decrease of energy of the substances involved. Some energy is transferred between the surroundings and the reactants of the reaction in the form of heat or light; thus the products of a reaction may have more or less energy than the reactants. A reaction is said to be exergonic if the final state is lower on the energy scale than the initial state; in the case of endergonic reactions the situation is the reverse. Chemical reactions are invariably not possible unless the reactants surmount an energy barrier known as the activation energy. The *speed* of a chemical reaction (at given temperature T) is related to the activation energy E, by the Boltzmann's population factor $e^{"E/kT}$ – that is the probability of molecule to have energy greater than or equal to E at the given temperature T. This exponential dependence of a reaction rate on temperature is known as the Arrhenius equation. The activation energy necessary for a chemical reaction can be in the form of thermal energy.
- In biology, energy is an attribute of all biological systems from the biosphere to the smallest living organism. Within an organism it is responsible for growth and development of a biological cell or an organelle of a biological organism. Energy is thus often said to be stored by cells in the structures of molecules of substances such as carbohydrates (including sugars), lipids, and proteins, which release energy when reacted with oxygen in respiration. In human terms, the human equivalent (H-e) (Human energy conversion) indicates, for a given amount of energy expenditure, the relative quantity of energy needed for human metabolism, assuming an average human energy expenditure of 12,500kJ per day and a basal metabolic rate of 80 watts. For example, if our bodies run (on average) at 80 watts, then a light bulb running at 100 watts is running at 1.25 human equivalents (100 χ 80) i.e. 1.25 H-e. For a difficult task of only a few seconds' duration, a person can put out thousands of watts, many times the 746 watts in one official horsepower. For tasks lasting a few minutes, a fit

human can generate perhaps 1,000 watts. For an activity that must be sustained for an hour, output drops to around 300; for an activity kept up all day, 150 watts is about the maximum. The human equivalent assists understanding of energy flows in physical and biological systems by expressing energy units in human terms: it provides a "feel" for the use of a given amount of energy.

- In geology, continental drift, mountain ranges, volcanoes, and earthquakes are phenomena that can be explained in terms of energy transformations in the Earth's interior, while meteorologicalphenomena like wind, rain, hail, snow, lightning, tornadoes and hurricanes, are all a result of energy transformations brought about by solar energy on the atmosphere of the planet Earth.
- In cosmology and astronomy the phenomena of stars, nova, supernova, quasars and gamma ray bursts are the universe's highest-output energy transformations of matter. All stellar phenomena (including solar activity) are driven by various kinds of energy transformations. Energy in such transformations is either from gravitational collapse of matter (usually molecular hydrogen) into various classes of astronomical objects (stars, black holes, etc.), or from nuclear fusion (of lighter elements, primarily hydrogen).

Energy transformations in the universe over time are characterised by various kinds of potential energy that has been available since the Big Bang, later being "released" (transformed to more active types of energy such as kinetic or radiant energy), when a triggering mechanism is available.

Familiar examples of such processes include nuclear decay, in which energy is released that was originally "stored" in heavy isotopes (such as uranium and thorium), by nucleosynthesis, a process ultimately using the gravitational potential energy released from the gravitational collapse of supernovae, to store energy in the creation of these heavy elements before they were incorporated into the solar system and the Earth. This energy is triggered and released in nuclear fission bombs. In a slower process, radioactive decay of these atoms in the core of the Earth releases heat. This thermal energy drives plate tectonics and may lift mountains, via orogenesis. This slow lifting represents a kind of gravitational potential energy storage of the thermal energy, which may be later released to active kinetic energy in landslides, after a triggering event. Earthquakes also release stored elastic potential energy in rocks, a store that has been produced ultimately from the same radioactive heat sources. Thus, according to present understanding, familiar events such as landslides and earthquakes release energy that has been stored as potential energy in the Earth's gravitational field or elastic strain (mechanical potential energy) in rocks. Prior to this, they represent release of energy that has been stored in heavy atoms since the collapse of long-destroyed supernova stars created these atoms.

In another similar chain of transformations beginning at the dawn of the universe, nuclear fusion of hydrogen in the Sun also releases another store of potential energy which was created at the time of the Big Bang. At that time, according to theory, space expanded and the universe cooled too rapidly for hydrogen to completely fuse into heavier elements.

This meant that hydrogen represents a store of potential energy that can be released by fusion. Such a fusion process is triggered by heat and pressure generated from gravitational collapse of hydrogen clouds when they produce stars, and some of the fusion energy is then transformed into sunlight.

Such sunlight from our Sun may again be stored as gravitational potential energy after it strikes the Earth, as (for example) water evaporates from oceans and is deposited upon mountains (where, after being released at a hydroelectric dam, it can be used to drive turbines or generators to produce electricity). Sunlight also drives many weather phenomena, save those generated by volcanic events. An example of a solar-mediated weather event is a hurricane, which occurs when large unstable areas of warm ocean, heated over months, give up some of their thermal energy suddenly to power a few days of violent air movement. Sunlight is also captured by plants as *chemical potential energy* in photosynthesis, when carbon dioxide and water (two low-energy compounds) are converted into the high-energy compounds carbohydrates, lipids, and proteins.

Plants also release oxygen during photosynthesis, which is utilised by living organisms as an electron acceptor, to release the energy of carbohydrates, lipids, and proteins. Release of the energy stored during photosynthesis as heat or light may be triggered suddenly by a spark, in a forest fire, or it may be made available more slowly for animal or human metabolism, when these molecules are ingested, and catabolism is triggered by enzyme action.

Through all of these transformation chains, potential energy stored at the time of the Big Bang is later released by intermediate events, sometimes being stored in a number of ways over time between releases, as more active energy. In all these events, one kind of energy is converted to other types of energy, including heat.

Distinction between Energy and Power

Although in everyday usage the terms *energy* and *power* are essentially synonyms, scientists and engineers distinguish between them. In its technical sense, power is not at all the same as energy, but is the *rate* at which energy is converted (or, equivalently, at which work is performed). Thus a hydroelectric plant, by allowing the water above the dam to pass through turbines, converts the water's potential energy into kinetic energy and ultimately into electric energy, whereas the amount of electric energy that is generated *per unit of time* is the electric power generated. The same amount of energy converted through a shorter period of time is more power over that shorter time.

Conservation of Energy

Energy is subject to the law of conservation of energy. According to this law, energy can neither be created (produced) nor destroyed by itself. It can only be transformed.

Most kinds of energy (with gravitational energy being a notable exception) are subject to strict local conservation laws as well. In this case, energy can only be exchanged between adjacent regions of space, and all observers agree as to the volumetric density of energy in any given space. There is also a global law of conservation of energy, stating that the total energy of the universe cannot change; this is a corollary of the local law, but not vice versa. Conservation of energy is the mathematical consequence of translational symmetry of time (that is, the indistinguishability of time intervals taken at different time) - see Noether's theorem.

According to Conservation of energy the total inflow of energy into a system must equal the total outflow of energy from the system, plus the change in the energy contained within the system.

This law is a fundamental principle of physics. It follows from the translational symmetry of time, a property of most phenomena below the cosmic scale that makes them independent of their locations on the time coordinate. Put differently, yesterday, today, and tomorrow are physically indistinguishable.

This is because energy is the quantity which is canonical conjugate to time. This mathematical entanglement of energy and time also results in the uncertainty principle - it is impossible to define the exact amount of energy during any definite time interval. The uncertainty principle should not be confused with energy conservation - rather it provides mathematical limits to which energy can in principle be defined and measured.

In quantum mechanics energy is expressed using the Hamiltonian operator. On any time scales, the uncertainty in the energy is by

$$\Delta E \Delta t \geq \frac{\hbar}{2}$$

which is similar in form to the Heisenberg uncertainty principle (but not really mathematically equivalent thereto, since H and t are not dynamically conjugate variables, neither in classical nor in quantum mechanics).

In particle physics, this inequality permits a qualitative understanding of virtual particles which carry momentum, exchange by which and with real particles, is responsible for the creation of all known fundamental forces (more accurately known as fundamental interactions). Virtual photons (which are simply lowest quantum mechanical energy state of photons) are also responsible for electrostatic interaction between electric charges (which results in Coulomb law), for spontaneous radiative decay of exited atomic and nuclear states, for the Casimir force, for van der Waals bond forces and some other observable phenomena.

Applications of the Concept of Energy

Energy is subject to a strict global conservation law; that is, whenever one measures (or calculates) the total energy of a system of particles whose interactions do not depend explicitly on time, it is found that the total energy of the system always remains constant.

- The total energy of a system can be subdivided and classified in various ways. For example, it is sometimes convenient to distinguish potential energy (which is a function of coordinates only) from kinetic energy (which is a function of coordinate time derivatives only). It may also be convenient to distinguish gravitational energy, electric energy, thermal energy, and other forms. These classifications overlap; for instance, thermal energy usually consists partly of kinetic and partly of potential energy.

- The *transfer* of energy can take various forms; familiar examples include work, heat flow, and advection, as discussed below.
- The word "energy" is also used outside of physics in many ways, which can lead to ambiguity and inconsistency. The vernacular terminology is not consistent with technical terminology. For example, while energy is always conserved (in the sense that the total energy does not change despite energy transformations), energy can be converted into a form, e.g., thermal energy, that cannot be utilised to perform work. When one talks about "conserving energy by driving less," one talks about conserving fossil fuels and preventing useful energy from being lost as heat. This usage of "conserve" differs from that of the law of conservation of energy.

In classical physics energy is considered a scalar quantity, the canonical conjugate to time. In special relativity energy is also a scalar (although not a Lorentz scalar but a time component of the energy-momentum 4-vector). In other words, energy is invariant with respect to rotations of space, but not invariant with respect to rotations of space-time (= boosts).

Energy Transfer

Because energy is strictly conserved and is also locally conserved (wherever it can be defined), it is important to remember that by the definition of energy the transfer of energy between the "system" and adjacent regions is work. A familiar example is *mechanical work*. In simple cases this is written as the following equation:

$$\Delta E = W \tag{1}$$

if there are no other energy-transfer processes involved. Here E is the amount of energy transferred, and W represents the work done on the system.

More generally, the energy transfer can be split into two categories:

$$\Delta E = W + Q \tag{2}$$

where Q represents the heat flow into the system.

There are other ways in which an open system can gain or lose energy. In chemical systems, energy can be added to a system by means of adding substances with different chemical potentials, which potentials are then extracted (both of these process are illustrated by fuelling an auto, a system which gains in energy thereby, without addition of either work or heat). Winding a clock would be adding energy to a mechanical system. These terms may be added to the above equation, or they can generally be subsumed into a quantity called "energy addition term E– which refers to *any* type of energy carried over the surface of a control volume or system volume. Examples may be seen above, and many others can be imagined (for example, the kinetic energy of a stream of particles entering a system, or energy from a laser beam adds to system energy, without either being either work-done or heat-added, in the classic senses).

$$\Delta E = W + Q + E \tag{3}$$

Where E in this general equation represents other additional advected energy terms not covered by work done on a system, or heat added to it.

Energy is also transferred from potential energy (E_p) to kinetic energy (E_k) and then back to potential energy constantly. This is referred to as conservation of energy. In this closed system, energy cannot be created or destroyed; therefore, the initial energy and the final energy will be equal to each other. This can be demonstrated by the following:

$$E_{pi} + E_{ki} = E_{pF} + E_{kF} \tag{4}$$

The equation can then be simplified further since $E_p = mgh$ (mass times acceleration due to gravity times the height) and $E_k = \frac{1}{2}mv^2$ (half mass times velocity squared). Then the total amount of energy can be found by adding $E_p + E_k = E_{total}$.

Energy and the Laws of Motion

In classical mechanics, energy is a conceptually and mathematically useful property, as it is a conserved quantity. Several formulations of mechanics have been developed using energy as a core concept.

The Hamiltonian

The total energy of a system is sometimes called the Hamiltonian, after William Rowan Hamilton. The classical equations of motion can be written in terms of the Hamiltonian, even for highly complex or abstract systems. These classical equations have remarkably direct analogs in nonrelativistic quantum mechanics.

The Lagrangian

Another energy-related concept is called the Lagrangian, after Joseph Louis Lagrange. This is even more fundamental than the Hamiltonian, and can be used to derive the equations of motion. It was invented in the context of classical mechanics, but is generally useful in modern physics. The Lagrangian is defined as the kinetic energy *minus* the potential energy.

Usually, the Lagrange formalism is mathematically more convenient than the Hamiltonian for non-conservative systems (such as systems with friction).

Noether's Theorem

Noether's (first) theorem (1918) states that any differentiable symmetry of the action of a physical system has a corresponding conservation law.

Noether's theorem has become a fundamental tool of modern theoretical physics and the calculus of variations. A generalisation of the seminal formulations on constants of motion in Lagrangian and Hamiltonian mechanics (1788 and 1833, respectively), it does not apply to systems that cannot be modelled with a Lagrangian; for example, dissipative systems with continuous symmetries need not have a corresponding conservation law.

Chapter 4

Position and its Derivatives

The *position* of a point particle is defined with respect to an arbitrary fixed reference point, O, in space, usually accompanied by a coordinate system, with the reference point located at the *origin* of the coordinate system. It is defined as the vector r from O to the particle. In general, the point particle need not be stationary relative to O, so r is a function of *t*, the time elapsed since an arbitrary initial time.

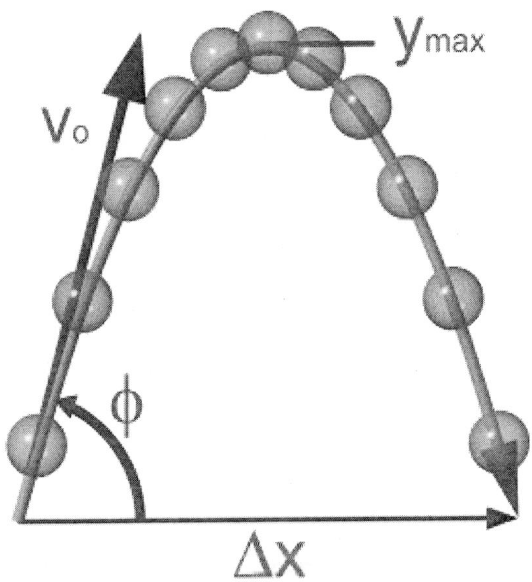

Figure: *The analysis of projectile motion is a part of classical mechanics.*

In pre-Einstein relativity (known as Galilean relativity), time is considered an absolute, i.e., the time interval between any given pair of events is the same for all observers. In addition to relying on absolute time, classical mechanics assumes Euclidean geometry for the structure of space.

Velocity and Speed

The *velocity*, or the rate of change of position with time, is defined as the derivative of the position with respect to time or

$$\mathbf{v} = \frac{d\mathbf{r}}{dt} .$$

In classical mechanics, velocities are directly additive and subtractive. For example, if one car travelling East at 60 km/h passes another car travelling East at 50 km/h, then from the perspective of the slower car, the faster car is travelling east at 60 • 50 = 10 km/h. Whereas, from the perspective of the faster car, the slower car is moving 10 km/h to the West. Velocities are directly additive as vector quantities; they must be dealt with using vector analysis.

Mathematically, if the velocity of the first object in the previous discussion is denoted by the vector u = *ud and* the velocity of the second object by the vector v = *v*e, where *u* is the speed of the first object, *v* is the speed of the second object, and d and e are unit vectors in the directions of motion of each particle respectively, then the velocity of the first object as seen by the second object is

$$\mathbf{u}' = \mathbf{u} - \mathbf{v}.$$

Similarly,

$$\mathbf{v}' = \mathbf{v} - \mathbf{u}.$$

When both objects are moving in the same direction, this equation can be simplified to

$$\mathbf{u}' = (u - v)\mathbf{d}.$$

Or, by ignoring direction, the difference can be given in terms of speed only:

$$u' = u - v.$$

Acceleration

The *acceleration*, or rate of change of velocity, is the derivative of the velocity with respect to time (the second derivative of the position with respect to time) or

$$\mathbf{a} = \frac{d\mathbf{v}}{dt}.$$

Acceleration can arise from a change with time of the magnitude of the velocity or of the direction of the velocity or both. If only the magnitude v of the velocity decreases, this is sometimes referred to as deceleration, but generally any change in the velocity with time, including deceleration, is simply referred to as acceleration.

Frames of Reference

While the position and velocity and acceleration of a particle can be referred to any observer in any state of motion, classical mechanics assumes the existence of a special family of reference frames in terms of which the mechanical laws of nature take a comparatively simple form. These special reference frames are called inertial frames. An inertial frame is such that when an object without any force interactions (an idealized situation) is viewed from it, it will appear either to be at rest or in a state of uniform motion in a straight line.

This is the fundamental definition of an inertial frame. They are characterized by the requirement that all forces entering the observer's physical laws originate in identifiable sources (charges, gravitational bodies, and so forth). A non-inertial reference frame is one accelerating with respect to an inertial one, and in such a non-inertial frame a particle is subject to acceleration by fictitious forces that enter the equations of motion solely as a result of its accelerated motion, and do not originate in identifiable sources. These fictitious forces are in addition to the real forces recognised in an inertial frame. A key concept of inertial frames is the method for identifying them. For practical purposes, reference frames that are unaccelerated with respect to the distant starsare regarded as good approximations to inertial frames.

Consider two reference frames S and S'. For observers in each of the reference frames an event has space-time coordinates of (x,y,z,t) in frame S and $(x2,y2,z2,t2)$ in frame $S2$. Assuming time is measured the same in all reference frames, and if we require $x = x'$ when $t = 0$, then the relation between the space-time coordinates of the same event observed from the reference frames $S2$ and S, which are moving at a relative velocity of u in the x direction is:

$$x2 = x\text{"}\ ut$$

$$y2 = y$$

$$z2 = z$$

$$t2 = t$$

This set of formulas defines a group transformation known as the Galilean transformation (informally, the *Galilean transform*). This group is a limiting case of the Poincaré group used in special relativity. The limiting case applies when the velocity u is very small compared to c, the speed of light.

The transformations have the following consequences:

- v2 = v " u (the velocity v2 of a particle from the perspective of $S2$ is slower by u than its velocity v from the perspective of S)
- a2 = a (the acceleration of a particle is the same in any inertial reference frame)
- F2 = F (the force on a particle is the same in any inertial reference frame)
- the speed of light is not a constant in classical mechanics, nor does the special position given to the speed of light in relativistic mechanics have a counterpart in classical mechanics.

For some problems, it is convenient to use rotating coordinates (reference frames). Thereby one can either keep a mapping to a convenient inertial frame, or introduce additionally a fictitious centrifugal force and Coriolis force.

Forces; Newton's Second Law

Newton was the first to mathematically express the relationship between force and momentum. Some physicists interpret Newton's second law of motion as a definition of force and mass, while others consider it to be a fundamental postulate, a law of nature. Either interpretation has the same mathematical consequences, historically known as "Newton's Second Law":

$$\mathbf{F} = \frac{\mathrm{d}\mathbf{p}}{\mathrm{d}t} = \frac{\mathrm{d}(m\mathbf{v})}{\mathrm{d}t}.$$

The quantity mv is called the (canonical) momentum. The net force on a particle is thus equal to rate change of momentum of the particle with time. Since the definition of acceleration is a = dv/dt, the second law can be written in the simplified and more familiar form:

$$\mathbf{F} = m\mathbf{a}.$$

So long as the force acting on a particle is known, Newton's second law is sufficient to describe the motion of a particle. Once independent relations for each force acting on a particle are available, they can be substituted into Newton's second law to obtain an ordinary differential equation, which is called the *equation of motion*. As an example, assume that friction is the only force acting on the particle, and that it may be modelled as a function of the velocity of the particle, for example:

$$\mathbf{F}_{\mathrm{R}} = -\lambda\mathbf{v},$$

where λ is a positive constant. Then the equation of motion is

$$-\lambda\mathbf{v} = m\mathbf{a} = m\frac{\mathrm{d}\mathbf{v}}{\mathrm{d}t}.$$

This can be integrated to obtain

$$\mathbf{v} = \mathbf{v}_0 e^{-\lambda t/m}$$

where v_0 is the initial velocity. This means that the velocity of this particle decays exponentially to zero as time progresses. In this case, an equivalent viewpoint is that the kinetic energy of the particle is absorbed by friction (which converts it to heat energy in accordance with the conservation of energy), slowing it down. This expression can be further integrated to obtain the position r of the particle as a function of time.

Important forces include the gravitational force and the Lorentz force for electromagnetism. In addition, Newton's third law can sometimes be used to deduce the forces acting on a particle: if it is known that particle A exerts a force F on another particle B, it follows that B must exert an equal and opposite reaction *force*, “F, on A. The strong form of Newton's third law requires that F and “F act along the line connecting A and B, while the weak form does not. Illustrations of the weak form of Newton's third law are often found for magnetic forces.

Work and Energy

If a constant force F is applied to a particle that achieves a displacement Δr, the *work done* by the force is defined as the scalar product of the force and displacement vectors:

$$W = \mathbf{F}\cdot\Delta\mathbf{r}.$$

More generally, if the force varies as a function of position as the particle moves from r_1 to r_2 along a path C, the work done on the particle is given by the line integral

$$W = \int_C \mathbf{F}(\mathbf{r})\cdot \mathrm{d}\mathbf{r}.$$

If the work done in moving the particle from r_1 to r_2 is the same no matter what path is taken, the force is said to be conservative. Gravity is a conservative force, as is the force due to an idealized spring, as given by Hooke's law. The force due to friction is non-conservative.

The kinetic energy E_k of a particle of mass m travelling at speed v is given by

$$E_k = \tfrac{1}{2}mv^2.$$

For extended objects composed of many particles, the kinetic energy of the composite body is the sum of the kinetic energies of the particles. The work-energy theorem states that for a particle of constant mass m the total work W done on the particle from position r_1 to r_2 is equal to the change in kinetic energy E_k of the particle:

$$W = \Delta E_k = E_{k,2} - E_{k,1} = \tfrac{1}{2}m\left(v_2^2 - v_1^2\right).$$

Conservative forces can be expressed as the gradient of a scalar function, known as the potential energy and denoted E_p:

$$\mathbf{F} = -\nabla E_p.$$

If all the forces acting on a particle are conservative, and E_p is the total potential energy (which is defined as a work of involved forces to rearrange mutual positions of bodies), obtained by summing the potential energies corresponding to each force

$$\mathbf{F}\cdot\Delta\mathbf{r} = -\nabla E_p\cdot\Delta\mathbf{r} = -\Delta E_p \Rightarrow -\Delta E_p = \Delta E_k \Rightarrow \Delta(E_k + E_p) = 0$$

This result is known as *conservation of energy* and states that the total energy,

$$\sum E = E_k + E_p.$$

is constant in time. It is often useful, because many commonly encountered forces are conservative.

Beyond Newton's Laws

Classical mechanics also includes descriptions of the complex motions of extended non-point like objects. Euler's laws provide extensions to Newton's laws in this area. The concepts of angular momentum rely on the same calculus used to describe one-dimensional motion. The rocket equation extends the notion of rate of change of an object's momentum to include the effects of an object "losing mass".

There are two important alternative formulations of classical mechanics: Lagrangian mechanics and Hamiltonian mechanics. These, and other modern formulations, usually bypass the concept of "force", instead referring to other physical quantities, such as energy, for describing mechanical systems.

The expressions given above for momentum and kinetic energy are only valid when there is no significant electromagnetic contribution.

In electromagnetism, Newton's second law for current-carrying wires breaks down unless one includes the electromagnetic field contribution to the momentum of the system as expressed by the Poynting vector divided by c^2, where c is the speed of light in free space.

History

Some Greek philosophers of antiquity, among them Aristotle, founder of Aristotelian physics, may have been the first to maintain the idea that "everything happens for a reason" and that theoretical principles can assist in the understanding of nature. While to a modern reader, many of these preserved ideas come forth as eminently reasonable, there is a conspicuous lack of both mathematical theory and controlled experiment, as we know it. These both turned out to be decisive factors in forming modern science, and they started out with classical mechanics.

The medieval "science of weights" (i.e., mechanics) owes much of its importance to the work of Jordanus de Nemore. In the *Elementa super demonstrationem ponderum*, he introduces the concept of "positional gravity" and the use of component forces.

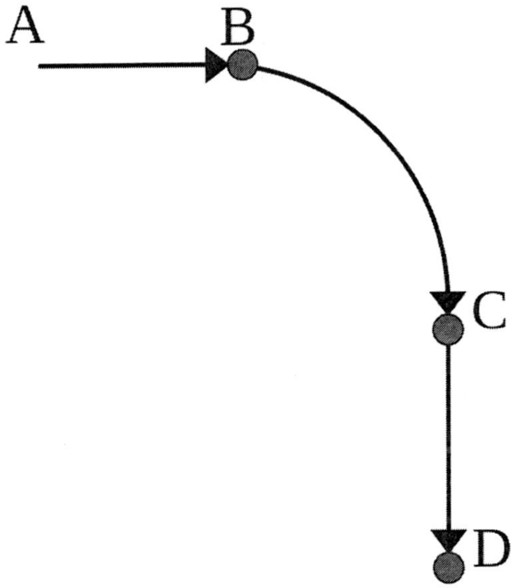

Figure: *Three stage Theory of impetus according to Albert of Saxony.*

The first published causal explanation of the motions of planets was Johannes Kepler's Astronomia nova published in 1609. He concluded, based on Tycho Brahe's observations of the orbit of Mars, that the orbits were ellipses. This break with ancient thought was happening around the same time that Galilei was proposing abstract mathematical laws for the motion of objects. He may (or may not) have performed the famous experiment of dropping two cannon balls of different weights from the tower of Pisa, showing that they both hit the ground at the same time.

The reality of this experiment is disputed, but, more importantly, he did carry out quantitative experiments by rolling balls on an inclined plane. His theory of accelerated motion derived from the results of such experiments, and forms a cornerstone of classical mechanics.

As foundation for his principles of natural philosophy, Newton proposed three laws of motion: the law of inertia, his second law of acceleration (mentioned above), and the law of action and reaction; and hence laid the foundations for classical mechanics.

Both Newton's second and third laws were given proper scientific and mathematical treatment in Newton's Philosophiζ Naturalis Principia Mathematica, which distinguishes them from earlier attempts at explaining similar phenomena, which were either incomplete, incorrect, or given little accurate mathematical expression. Newton also enunciated the principles of conservation of momentum and angular momentum.

In Mechanics, Newton was also the first to provide the first correct scientific and mathematical formulation of gravity in Newton's law of universal gravitation. The combination of Newton's laws of motion and gravitation provide the fullest and most accurate description of classical mechanics. He demonstrated that these laws apply to everyday objects as well as to celestial objects. In particular, he obtained a theoretical explanation of Kepler's laws of motion of the planets.

Newton previously invented the calculus, of mathematics, and used it to perform the mathematical calculations. For acceptability, his book, the Principia, was formulated entirely in terms of the long established geometric methods, which were soon to be eclipsed by his calculus. However it was Leibniz who developed the notation of the derivative and integral preferred today. Newton, and most of his contemporaries, with the notable exception of Huygens, worked on the assumption that classical mechanics would be able to explain all phenomena, including light, in the form of geometric optics. Even when discovering the so-called Newton's rings (a wave interference phenomenon) his explanation remained with his own corpuscular theory of light.

After Newton, classical mechanics became a principal field of study in mathematics as well as physics. After Newton there were several re-formulations which progressively allowed a solution to be found to a far greater number of problems. The first notable re-formulation was in 1788 by Joseph Louis Lagrange. Lagrangian mechanics was in turn re-formulated in 1833 by William Rowan Hamilton.

Some difficulties were discovered in the late 19th century that could only be resolved by more modern physics. Some of these difficulties related to compatibility with electromagnetic theory, and the famous Michelson-Morley experiment.

The resolution of these problems led to the special theory of relativity, often included in the term classical mechanics. A second set of difficulties were related to thermodynamics.

When combined with thermodynamics, classical mechanics leads to the Gibbs paradox of classical statistical mechanics, in which entropy is not a well-defined quantity. Black-body radiation was not explained without the introduction of quanta.

As experiments reached the atomic level, classical mechanics failed to explain, even approximately, such basic things as the energy levels and sizes of atoms and the photo-electric effect. The effort at resolving these problems led to the development of quantum mechanics.

Since the end of the 20th century, the place of classical mechanics in physics has been no longer that of an independent theory.

Emphasis has shifted to understanding the fundamental forces of nature as in the Standard model and its more modern extensions into a unified theory of everything. Classical mechanics is a theory for the study of the motion of non-quantum mechanical, low-energy particles in weak gravitational fields. In the 21st century classical mechanics has been extended into the complex domain and complex classical mechanics exhibits behaviours very similar to quantum mechanics.

Limits of Validity

Many branches of classical mechanics are simplifications or approximations of more accurate forms; two of the most accurate being general relativity and relativistic statistical mechanics. Geometric optics is an approximation to thequantum theory of light, and does not have a superior "classical" form.

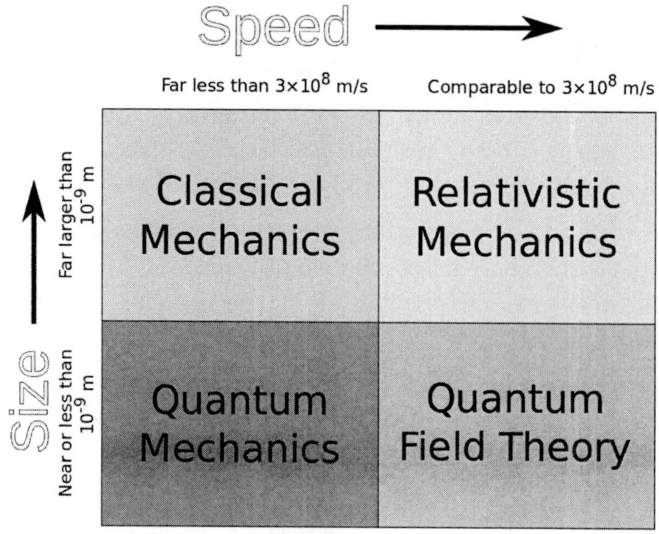

Figure: *Domain of validity for Classical Mechanics*

The Newtonian Approximation to Special Relativity

In special relativity, the momentum of a particle is given by

$$\mathbf{p} = \frac{m\mathbf{v}}{\sqrt{1-(v^2/c^2)}},$$

where m is the particle's mass, v its velocity, and c is the speed of light.

If v is very small compared to c, v^2/c^2 is approximately zero, and so

$$\mathbf{p} \approx m\mathbf{v}.$$

Thus the Newtonian equation p = mv is an approximation of the relativistic equation for bodies moving with low speeds compared to the speed of light.

For example, the relativistic cyclotron frequency of a cyclotron, gyrotron, or high voltage magnetron is given by

$$f = f_c \frac{m_0}{m_0 + T/c^2},$$

where f_c is the classical frequency of an electron (or other charged particle) with kinetic energy T and (rest) mass m_0 circling in a magnetic field. The (rest) mass of an electron is 511 keV. So the frequency correction is 1% for a magnetic vacuum tube with a 5.11 kV direct current accelerating voltage.

The Classical Approximation to Quantum Mechanics

The ray approximation of classical mechanics breaks down when the de Broglie wavelength is not much smaller than other dimensions of the system. For non-relativistic particles, this wavelength is

$$\lambda = \frac{h}{p}$$

where h is Planck's constant and p is the momentum.

Again, this happens with electrons before it happens with heavier particles. For example, the electrons used by Clinton Davisson and Lester Germer in 1927, accelerated by 54 volts, had a wave length of 0.167 nm, which was long enough to exhibit a single diffraction side lobe when reflecting from the face of a nickel crystal with atomic spacing of 0.215 nm. With a larger vacuum chamber, it would seem relatively easy to increase the angular resolution from around a radian to a milliradian and see quantum diffraction from the periodic patterns of integrated circuit computer memory.

More practical examples of the failure of classical mechanics on an engineering scale are conduction by quantum tunnelling in tunnel diodes and very narrow transistor gates in integrated circuits. Classical mechanics is the same extreme high frequency approximation as geometric optics. It is more often accurate because it describes particles and bodies with rest mass. These have more momentum and therefore shorter De Broglie wavelengths than massless particles, such as light, with the same kinetic energies.

Branches

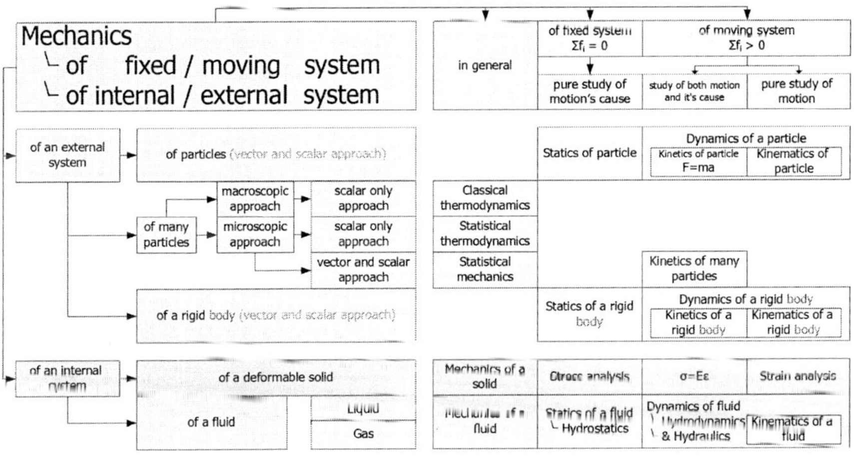

Figure: Branches of mechanics

Classical mechanics was traditionally divided into three main branches:

- Statics, the study of equilibrium and its relation to forces
- Dynamics, the study of motion and its relation to forces
- Kinematics, dealing with the implications of observed motions without regard for circumstances causing them

Another division is based on the choice of mathematical formalism:

- Newtonian mechanics
- Lagrangian mechanics
- Hamiltonian mechanics

Alternatively, a division can be made by region of application:

- Celestial mechanics, relating to stars, planets and other celestial bodies
- Continuum mechanics, for materials which are modelled as a continuum, e.g., solids and fluids(i.e., liquids and gases).
- Relativistic mechanics (i.e. including the special and general theories of relativity), for bodies whose speed is close to the speed of light.
- Statistical mechanics, which provides a framework for relating the microscopic properties of individual atoms and molecules to the macroscopic or bulk thermodynamic properties of materials.

Ground State

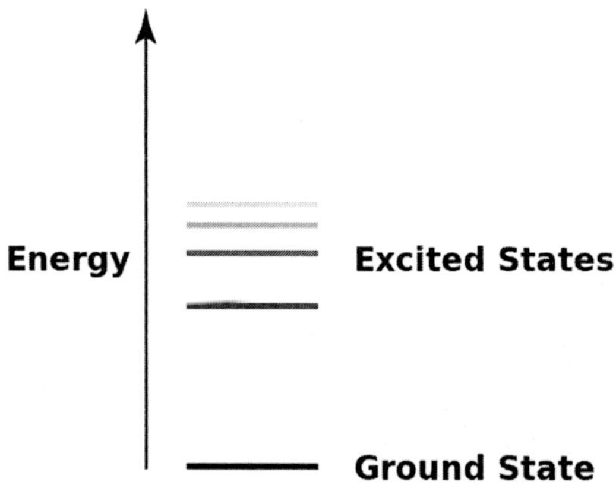

Figure: *Energy levels for an electron in an atom: ground state and excited states. After absorbing energy, an electron may jump from the ground state to a higher energy excited state.*

The ground state of a quantum mechanical system is its lowest-energy state; the energy of the ground state is known as the zero-point energy of the system. An excited state is any state with energy greater than the ground state. The ground state of a quantum field theory is usually called the vacuum state or the vacuum.

If more than one ground state exists, they are said to be degenerate. Many systems have degenerate ground states. It turns out that degeneracy occurs whenever a nontrivial unitary operator commutes with the Hamiltonian of the system.

According to the third law of thermodynamics, a system at absolute zero temperature exists in its ground state; thus, its entropy is determined by the degeneracy of the ground state. Many systems, such as a perfect crystal lattice, have a unique ground state and therefore have zero entropy at absolute zero because the logarithm of 1 is 0. It is also possible for the highest excited state to have absolute zero temperature for systems that exhibit negative temperature.

Examples

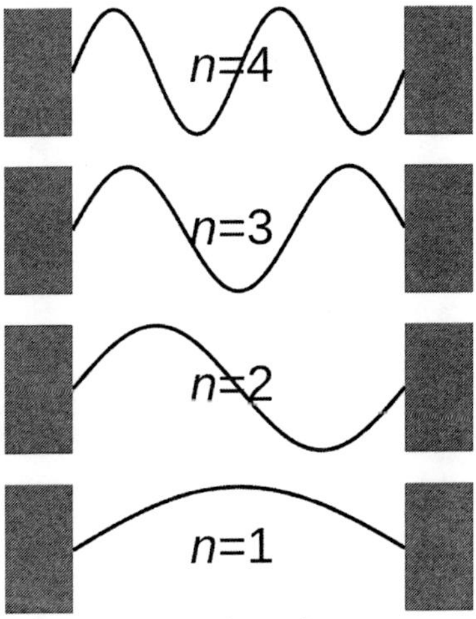

Figure: *Initial wave functions for the first four states of a one-dimensional particle in a box*

- The wave function of the ground state of a particle in a one-dimensional well is a half-period sine wave which goes to zero at the two edges of the well. The energy of the particle is given by $\frac{h^2 n^2}{8mL^2}$, where h is the Planck constant, m is the mass of the particle, n is the energy state ($n = 1$ corresponds to the ground-state energy), and L is the width of the well.

- The wave function of the ground state of a hydrogen atom is a spherically-symmetric distribution centred on the nucleus, which is largest at the centre and reduces exponentially at larger distances. The electron is most likely to be found at a distance from the nucleus equal to the Bohr radius. This function is known as the 1s atomic orbital. For hydrogen (H), an electron in the ground state has energy "13.6 eV, relative to 0.0 eV when the H atom is ionized, e.g. the electron is completely removed.

Atomic Theory

In chemistry and physics, atomic theory is a theory of the nature of matter, which states that matter is composed of discrete units called atoms, as opposed to the obsolete notion that matter could be divided into any arbitrarily small quantity. It began as a philosophical concept in ancient Greece (Democritus) and India and entered the scientific mainstream in the early 19th century when discoveries in the field of chemistry showed that matter did indeed behave as if it were made up of particles.

The word "atom" (from the ancient Greek adjective *atomos*, 'indivisible') was applied to the basic particle that constituted a chemical element, because the chemists of the era believed that these were the fundamental particles of matter. However, around the turn of the 20th century, through various experiments with electromagnetism and radioactivity, physicists discovered that the so-called "indivisible atom" was actually a conglomerate of various subatomic particles (chiefly, electrons, protons and neutrons) which can exist separately from each other. In fact, in certain extreme environments such as neutron stars, extreme temperature and pressure prevents atoms from existing at all. Since atoms were found to be actually divisible, physicists later invented the term "elementary particles" to describe indivisible particles. The field of science which studies subatomic particles is particle physics, and it is in this field that physicists hope to discover the true fundamental nature of matter.

Modern Atomic Theory

Earliest Empirical Evidence: Near the end of the 18th century, two laws about chemical reactions emerged without referring to the notion of an atomic theory. The first was the law of conservation of mass, formulated by Antoine Lavoisier in 1789, which states that the total mass in a chemical reaction remains constant (that is, the reactants have the same mass as the products). The second was the law of definite proportions. First proven by the French chemist Joseph Louis Proust in 1799, this law states that if a compound is broken down into its constituent elements, then the masses of the constituents will always have the same proportions, regardless of the quantity or source of the original substance.

John Dalton studied and expanded upon this previous work and developed the law of multiple proportions: if two elements came together form more than one compound, then the ratios of the masses of the second element which combine with a fixed mass of the first element will be ratios of small integers.

For instance, Proust had studied tin oxides and found that their masses were either 88.1% tin and 11.9% oxygen or 78.7% tin and 21.3% oxygen (these were tin(II) oxide and tin dioxide respectively). Dalton noted from these percentages that 100g of tin will combine either with 13.5g or 27g of oxygen; 13.5 and 27 form a ratio of 1:2. Dalton found an atomic theory of matter could elegantly explain this common pattern in chemistry - in the case of Proust's tin oxides, one tin atom will combine with either one or two oxygen atoms.

Dalton also believed atomic theory could explain why water absorbed different gases in different proportions: for example, he found that water absorbed carbon dioxide far better than it absorbed nitrogen. Dalton hypothesized this was due to the differences in mass and complexity of the gases' respective particles. Indeed, carbon dioxide molecules (CO_2) are heavier and larger than nitrogen molecules (N_2).

Dalton proposed that each chemical element is composed of atoms of a single, unique type, and though they cannot be altered or destroyed by chemical means, they can combine to form more complex structures (chemical compounds). This marked the first truly scientific theory of the atom, since Dalton reached his conclusions by experimentation and examination of the results in an empirical fashion.

In 1803 Dalton orally presented his first list of relative atomic weights for a number of substances. This paper was published in 1805, but he did not discuss there exactly how he obtained these figures. The method was first revealed in 1807 by his acquaintance Thomas Thomson, in the third edition of Thomson's textbook, *A System of Chemistry*. Finally, Dalton published a full account in his own textbook, A *New System of Chemical Philosophy*, 1808 and 1810.

Dalton estimated the atomic weights according to the mass ratios in which they combined, with hydrogen being the basic unit. However, Dalton did not conceive that with some elements atoms exist in molecules — e.g. pure oxygen exists as O_2. He also mistakenly believed that the simplest compound between any two elements is always one atom of each (so he thought water was HO, not H_2O). This, in addition to the crudity of his equipment, resulted in his table being highly flawed. For instance, he believed oxygen atoms were 5.5 times heavier than hydrogen atoms, because in water he measured 5.5 grams of oxygen for every 1 gram of hydrogen and believed the formula for water was HO (an oxygen atom is actually 16 times heavier than a hydrogen atom).

The flaw in Dalton's theory was corrected in 1811 by Amedeo Avogadro. Avogadro had proposed that equal volumes of any two gases, at equal temperature and pressure, contain equal numbers of molecules (in other words, the mass of a gas's particles does not affect its volume). Avogadro's law allowed him to deduce the diatomic nature of numerous gases by studying the volumes at which they reacted. For instance: since two litres of hydrogen will react with just one litre of oxygen to produce two litres of water vapour (at constant pressure and temperature), it meant a single oxygen molecule splits in two in order to form two particles of water. Thus, Avogadro was able to offer more accurate estimates of the atomic mass of oxygen and various other elements, and firmly established the distinction between molecules and atoms.

In 1827, the British botanist Robert Brown observed that dust particles inside pollen grains floating in water constantly jiggled about for no apparent reason. In 1905, Albert Einstein theorized that this Brownian motion was caused by the water molecules continuously knocking the grains about, and developed a hypothetical mathematical model to describe it. This model was validated experimentally in 1908 by French physicist Jean Perrin, thus providing additional validation for particle theory (and by extension atomic theory).

Discovery of Subatomic Particles

Atoms were thought to be the smallest possible division of matter until 1897 when J.J. Thomson discovered the electron through his work on cathode rays. A Crookes tubei s a sealed glass container in which two electrodes are separated by a vacuum. When a voltage is applied across the electrodes, cathode rays are generated, creating a glowing patch where they strike the glass at the opposite end of the tube. Through experimentation, Thomson discovered that the rays could be deflected by an electric field (in addition to magnetic fields, which was already known). He concluded that these rays, rather than being a form of light, were composed of very light negatively charged particles he called "corpuscles" (they would later be renamed electrons by other scientists).

Thomson believed that the corpuscles emerged from the molecules of gas around the cathode. He thus concluded that atoms were divisible, and that the corpuscles were their building blocks. To explain the overall neutral charge of the atom, he proposed that the corpuscles were distributed in a uniform sea of positive charge; this was the plum pudding model as the electrons were embedded in the positive charge like plums in a plum pudding (although in Thomson's model they were not stationary). Thomson's illustration of the Crookes tube by which he proved the particle nature of cathode rays. Cathode rays were emitted from the cathode C, sharpened to a beam by slits A and B, then passed through the electric field generated between plates D and E. When the cathode ray (blue line) passed through the electric field (yellow), it was deflected.

Discovery of the Nucleus

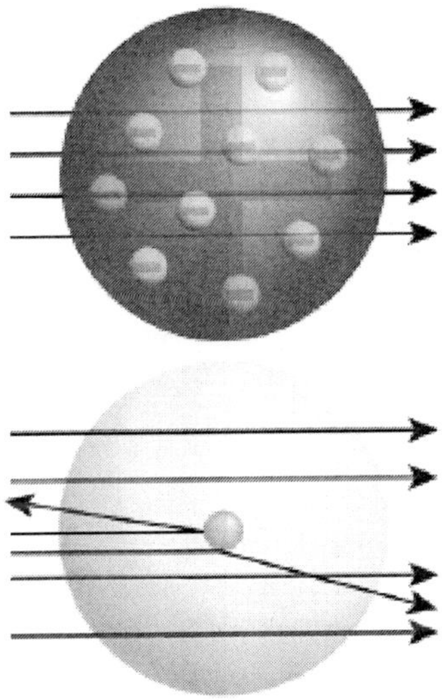

Figure: *The gold foil experiment*

Top. Expected results: alpha particles passing through the plum pudding model of the atom with negligible deflection.

Bottom: Observed results: a small portion of the particles were deflected by the concentrated positive charge of the nucleus.

Thomson's plum pudding model was disproved in 1909 by one of his former students, Ernest Rutherford, who discovered that most of the mass and positive charge of an atom is concentrated in a very small fraction of its volume, which he assumed to be at the very centre.

In the gold foil experiment, Hans Geiger and Ernest Marsden (colleagues of Rutherford working at his behest) shot alpha particles at a thin sheet of gold, measuring their deflection with a fluorescent screen. Given the very small mass of the electrons, the high momentum of the alpha particles and the unconcentrated distribution of positive charge of the plum pudding model, the experimenters expected all the alpha particles to pass through the gold sheet without significant deflection. To their astonishment, a small fraction of the alpha particles experienced heavy deflection.

This led Rutherford to propose a planetary model in which a cloud of electrons surrounded a small, compact nucleus of positive charge. Only such a concentration of charge could produce the electric field strong enough to cause the heavy deflection.

First Steps Toward a Quantum Physical Model of the Atom

The planetary model of the atom had two significant shortcomings. The first is that, unlike planets orbiting a sun, electrons are charged particles. An accelerating electric charge is known to emit electromagnetic waves according to the Larmor formulain classical electromagnetism; an orbiting charge should steadily lose energy and spiral toward the nucleus, colliding with it in a small fraction of a second. The second problem was that the planetary model could not explain the highly peaked emission and absorption spectra of atoms that were observed.

Quantum theory revolutionized physics at the beginning of the 20th century, when Max Planck and Albert Einstein postulated that light energy is emitted or absorbed in discrete amounts known as quanta (singular, *quantum*). In 1913, Niels Bohr incorporated this idea into his Bohr model of the atom, in which an electron could only orbit the nucleus in particular circular orbits with fixed angular momentum and energy, its distance from the nucleus (i.e., their radii) being proportional to its energy.

Under this model an electron could not spiral into the nucleus because it could not lose energy in a continuous manner; instead, it could only make instantaneous "quantum leaps" between the fixed energy levels. When this occurred, light was emitted or absorbed at a frequency proportional to the change in energy (hence the absorption and emission of light in discrete spectra).

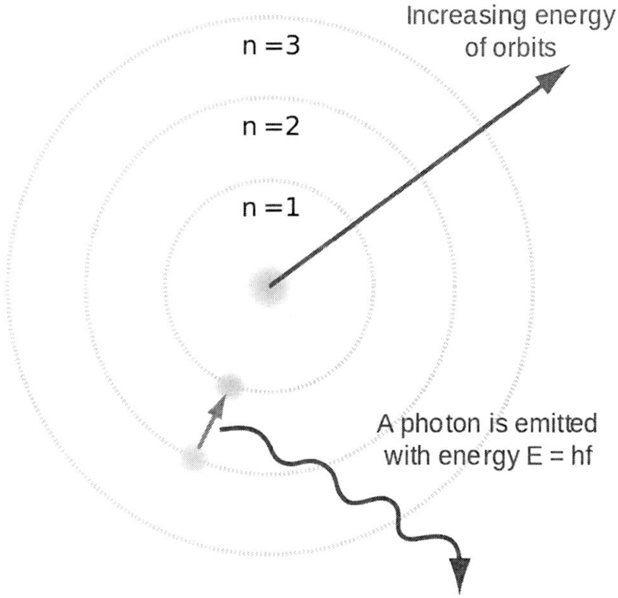

Figure: *The Bohr model of the atom*

Bohr's model was not perfect. It could only predict the spectral lines of hydrogen; it couldn't predict those of multielectron atoms. Worse still, as spectrographic technology improved, additional spectral lines in hydrogen were observed which Bohr's model couldn't explain. In 1916, Arnold Sommerfeld added elliptical orbits to the Bohr model to explain the extra emission lines, but this made the model very difficult to use, and it still couldn't explain more complex atoms.

Discovery of Isotopes

While experimenting with the products of radioactive decay, in 1913 radiochemist Frederick Soddy discovered that there appeared to be more than one element at each position on the periodic table. The term isotope was coined by Margaret Todd as a suitable name for these elements. That same year, J.J. Thomson conducted an experiment in which he channelled a stream of neon ions through magnetic and electric fields, striking a photographic plate at the other end. He observed two glowing patches on the plate, which suggested two different deflection trajectories. Thomson concluded this was because some of the neon ions had a different mass. The nature of this differing mass would later be explained by the discovery of neutrons in 1932.

Discovery of Nuclear Particles

In 1917 Rutherford bombarded nitrogen gas with alpha particles and observed hydrogen nuclei being emitted from the gas (Rutherford recognised these, because he had previously obtained them bombarding hydrogen with alpha particles, and observing hydrogen nuclei in the products). Rutherford concluded that the hydrogen nuclei emerged from the nuclei of the nitrogen atoms themselves (in effect, he had split a nitrogen).

From his own work and the work of his students Bohr and Henry Moseley, Rutherford knew that positive charge of any atom could always be equated to that of an integer number of hydrogen nuclei. This, coupled with the facts that hydrogen was the lightest element known and that the atomic mass of many elements was roughly equivalent to an integer number of hydrogen atoms (Prout's law), led him to conclude that hydrogen nuclei were singular particles and a basic constituent of all atomic nuclei. He named such particles protons. Further experimentation by Rutherford found that the nuclear mass of most atoms exceeded that of the protons it possessed; he speculated that this surplus mass was composed of hitherto unknown neutrally charged particles, which were tentatively dubbed "neutrons". Other scientists speculated that the nucleus perhaps contained "nuclear electrons" that neutralized the charges of some nuclear protons (this turned out not to be the case).

In 1928, Walter Bothe observed that beryllium emitted a highly penetrating, electrically neutral radiation when bombarded with alpha particles. It was later discovered that this radiation could knock hydrogen atoms out of paraffin wax. Initially it was thought to be high-energy gamma radiation, since gamma radiation had a similar effect on electrons in metals, but James Chadwick found that the ionization effect was too strong for it to be due to electromagnetic radiation, so long as energy and momentum were conserved in the interaction. In 1932, Chadwick exposed various elements, such as hydrogen and nitrogen, to the mysterious "beryllium radiation", and by measuring the energies of the recoiling charged particles, he deduced that the radiation was actually composed of electrically neutral particles which could not be massless like the gamma ray, but instead were required to have a mass similar to that of a proton. Chadwick now claimed these particles as Rutherford's neutrons. For his discovery of the neutron, Chadwick received the Nobel Prize in 1935.

Quantum Physical Models of the Atom

In 1924, Louis de Broglie proposed that all moving particles — particularly subatomic particles such as electrons — exhibit a degree of wave-like behaviour. Erwin Schrödinger, fascinated by this idea, explored whether or not the movement of an electron in an atom could be better explained as a wave rather than as a particle. Schrödinger's equation, published in 1926, describes an electron as a wavefunction instead of as a point particle. This approach elegantly predicted many of the spectral phenomena that Bohr's model failed to explain. Although this concept was mathematically convenient, it was difficult to visualize, and faced opposition. One of its critics, Max Born, proposed instead that Schrödinger's wavefunction described not the electron but rather all its possible states, and thus could be used to calculate the probability of finding an electron at any given location around the nucleus. This reconciled the two opposing theories of particle versus wave electrons and the idea of wave-particle duality was introduced. This theory stated that the electron may exhibit the properties of both a wave and a particle. For example, it can be refracted like a wave, and has mass like a particle.

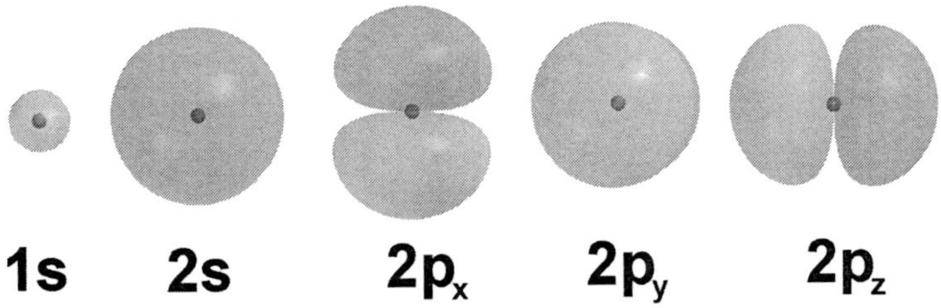

***Figure:** The five filled atomic orbitals of a neon atom separated and arranged in order of increasing energy from left to right, with the last three orbitals being equal in energy. Each orbital holds up to two electrons, which most probably exist in the zones represented by the coloured bubbles. Each electron is equally present in both orbital zones, shown here by colour only to highlight the different wave phase.*

A consequence of describing electrons as waveforms is that it is mathematically impossible to simultaneously derive the position and momentum of an electron; this became known as the Heisenberg uncertainty principle after the theoretical physicist Werner Heisenberg, who first described it and published it in 1927. This invalidated Bohr's model, with its neat, clearly defined circular orbits. The modern model of the atom describes the positions of electrons in an atom in terms of probabilities. An electron can potentially be found at any distance from the nucleus, but, depending on its energy level, exists more frequently in certain regions around the nucleus than others; this pattern is referred to as its atomic orbital. The orbitals come in a variety of shapes-sphere, dumbbell, torus, etc.-with the nucleus in the middle.

Corpuscular Theory of Light

In optics, corpuscular theory of light, set forward by Sir Isaac Newton, states that light is made up of small discrete particles called "corpuscles" (little particles) which travel in a straight line with a finite velocity and possess kinetic energy.

Newton's theory remained in force for more than 100 years and took precedence over Huygens' wave front theory, partly because of Newton's great prestige. However when the corpuscular theory failed to adequately explain the diffraction, interference and polarization of light it was abandoned in favour of Huygen's wave theory.

Newton's corpuscular theory was an elaboration of his view of reality as interactions of material points through forces. Note Albert Einstein's description of Newton's conception of physical reality:

[Newton's] physical reality is characterised by concepts of space, time, the material point and force (interaction between material points). Physical events are to be thought of as movements according to law of material points in space. The material point is the only representative of reality in so far as it is subject to change. The concept of the material point is obviously due to observable bodies; one conceived of the material point on the analogy of movable bodies by omitting characteristics of extension, form, spatial locality, and all their 'inner' qualities, retaining only inertia, translation, and the additional concept of force.

Chapter 5

Angular Momentum

In physics, angular momentum, moment of momentum, or rotational momentum is a conserved vector quantity that can be used to describe the overall state of a physical system. The angular momentum L of a particle with respect to some point of origin is

$$\mathbf{L} = \mathbf{r} \times \mathbf{p} = \mathbf{r} \times m\mathbf{v},$$

where r is the particle's position from the origin, p = mv is its linear momentum, and × denotes the cross product.

The angular momentum of a system of particles (e.g. a rigid body) is the sum of angular momenta of the individual particles. For a rigid body rotating around an axis of symmetry (e.g. the fins of a ceiling fan), the angular momentum can be expressed as the product of the body's moment of inertia I (a measure of an object's resistance to changes in its rotation rate) and its angular velocity ω:

$$\mathrm{L} = I\omega$$

In this way, angular momentum is sometimes described as the rotational analog of linear momentum.

Angular momentum is conserved in a system where there is no net external torque, and its conservation helps explain many diverse phenomena. For example, the increase in rotational speed of a spinning figure skater as the skater's arms are contracted is a consequence of conservation of angular momentum. The very high rotational rates of neutron stars can also be explained in terms of angular momentum conservation. Moreover, angular momentum conservation has numerous applications in physics and engineering (e.g. the gyrocompass).

Angular Momentum in Classical Mechanics

Definition

The angular momentum L of a particle about a given origin is defined as:

$$\mathbf{L} = \mathbf{r} \times \mathbf{p}$$

where r is the position vector of the particle relative to the origin, p is the linear momentum of the particle, and × denotes the cross product.

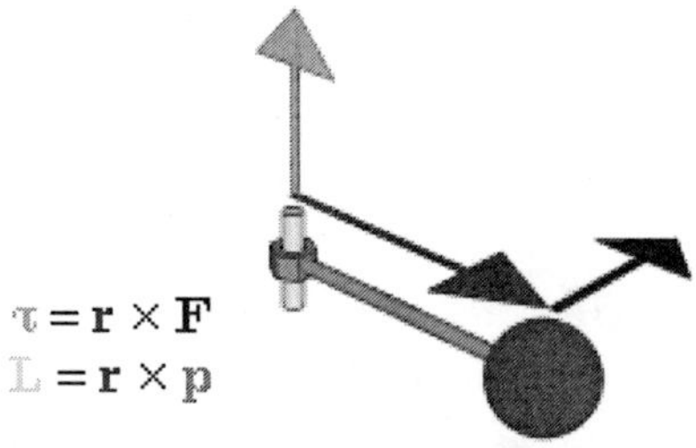

Figure: *Relationship between force (F), torque (τ), momentum (p), and angular momentum (L) vectors in a rotating system*

As seen from the definition, the derived SI units of angular momentum are newton metre seconds ($N \cdot m \cdot s$ or $kg \cdot m^2 s^{-1}$) or joule seconds. Because of the cross product, L is a pseudovector perpendicular to both the radial vector r and the momentum vector p and it is assigned a sign by the right-hand rule.

For an object with a fixed mass that is rotating about a fixed symmetry axis, the angular momentum is expressed as the product of the moment of inertia of the object and its angular velocity vector:

$$\mathbf{L} = I\omega$$

where I is the moment of inertia of the object (in general, a tensor quantity), and ω is the angular velocity.

The angular momentum of a particle or rigid body in rectilinear motion (pure translation) is a vector with constant magnitude and direction. If the path of the particle or rigid body passes through the given origin, its angular momentum is zero.

Angular momentum is also known as moment of momentum.

Angular Momentum of a Collection of Particles

If a system consists of several particles, the total angular momentum about a point can be obtained by adding (or integrating) all the angular momenta of the constituent particles.

Angular Momentum Simplified Using the Centre of Mass

It is very often convenient to consider the angular momentum of a collection of particles about their centre of mass, since this simplifies the mathematics considerably. The angular momentum of a collection of particles is the sum of the angular momentum of each particle:

$$\mathbf{L} = \sum_i \mathbf{R}_i \times m_i \mathbf{V}_i$$

where R_i is the position vector of particle i from the reference point, m_i is its mass, and V_i is its velocity. The centre of mass is defined by:

$$\mathbf{R} = \frac{1}{M}\sum_i m_i \mathbf{R}_i$$

where the total mass of all particles is given by

$$M = \sum_i m_i.$$

It follows that the velocity of the centre of mass is

$$\mathbf{V} = \frac{1}{M}\sum_i m_i \mathbf{V}_i.$$

If we define r_i as the displacement of particle i from the centre of mass, and v_i as the velocity of particle i with respect to the centre of mass, then we have

$$\mathbf{R}_i = \mathbf{R} + \mathbf{r}_i \quad \text{and} \quad \mathbf{V}_i = \mathbf{V} + \mathbf{v}_i$$

and also

$$\sum_i m_i \mathbf{r}_i = 0 \quad \text{and} \quad \sum_i m_i \mathbf{v}_i = 0$$

so that the total angular momentum with respect to the centre is

$$\mathbf{L} = \sum_i (\mathbf{R} + \mathbf{r}_i) \times m_i (\mathbf{V} + \mathbf{v}_i) = (\mathbf{R} \times M\mathbf{V}) + \left(\sum_i \mathbf{r}_i \times m_i \mathbf{v}_i\right).$$

The first term is just the angular momentum of the centre of mass. It is the same angular momentum one would obtain if there were just one particle of mass M moving at velocity V located at the centre of mass. The second term is the angular momentum that is the result of the particles moving relative to their centre of mass. This second term can be even further simplified if the particles form a rigid body, in which case it is the product of moment of inertia and angular velocity of the spinning motion (as above). The same result is true if the discrete point masses discussed above are replaced by a continuous distribution of matter.

Fixed Axis of Rotation

For many applications where one is only concerned about rotation around one axis, it is sufficient to discard the pseudovector nature of angular momentum, and treat it like a scalar where it is positive when it corresponds to a counter-clockwise rotation, and negative clockwise.

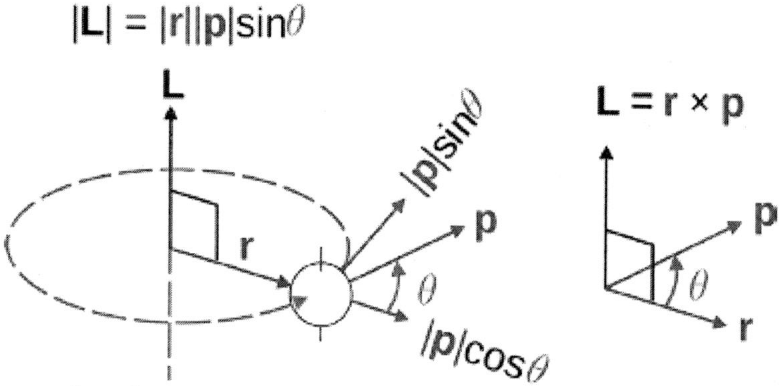

Figure: *Angular momentum in terms of scalar and vector components.*

To do this, just take the definition of the cross product and discard the unit vector, so that angular momentum becomes:

$$L = |\mathbf{r}|\,|\mathbf{p}|\sin\theta_{r,p}$$

where $\theta_{r,p}$ is the angle between r and p measured from r to p; an important distinction because without it, the sign of the cross product would be meaningless. From the above, it is possible to reformulate the definition to either of the following:

$$L = \pm|\mathbf{p}|\,|\mathbf{r}_\perp|$$

where $\mathbf{r}_\perp$ is called the *lever arm distance* to p.

The easiest way to conceptualize this is to consider the lever arm distance to be the distance from the origin to the line that p travels along. With this definition, it is necessary to consider the direction of p (pointed clockwise or counter-clockwise) to figure out the sign of L. Equivalently:

$$L = \pm|\mathbf{r}|\,|\mathbf{p}_\perp|$$

where $\mathbf{p}_\perp$ is the component of p that is perpendicular to r. As above, the sign is decided based on the sense of rotation.

For an object with a fixed mass that is rotating about a fixed symmetry axis, the angular momentum is expressed as the product of the moment of inertia of the object and its angular velocity vector:

$$\mathbf{L} = I\omega$$

where I is the moment of inertia of the object (in general, a tensor quantity) and ω is the angular velocity.

It is a misconception that angular momentum is always about the same axis as angular velocity. Sometime this may not be possible, in these cases the angular momentum component along the axis of rotation is the product of angular velocity and moment of inertia about the given axis of rotation.

As the kinetic energy K of a massive rotating body is given by

$$K = I\omega^2 / 2$$

it is proportional to the square of the angular velocity.

Conservation of Angular Momentum

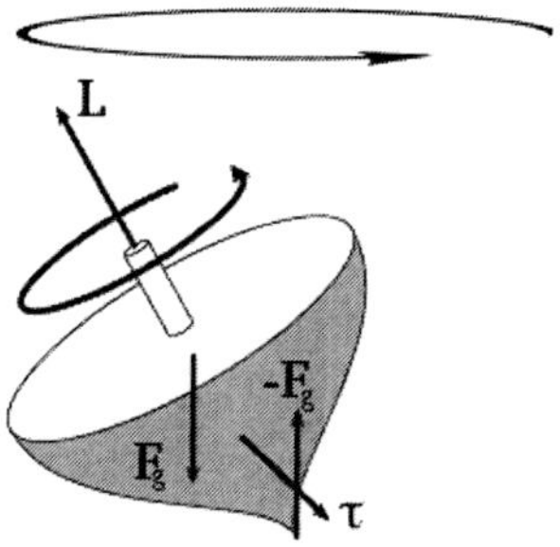

The torque caused by the two opposing forces F_g and $-F_g$ causes a change in the angular momentum L in the direction of that torque (since torque is the time derivative of angular momentum). This causes the top to precess.

In a closed system angular momentum is constant. This conservation law mathematically follows from continuous directional symmetry of space (no direction in space is any different from any other direction). See Noether's theorem.

The time derivative of angular momentum is called torque:

$$\tau = \frac{d\mathbf{L}}{dt} = \frac{d\mathbf{r}}{dt} \times \mathbf{p} + \mathbf{r} \times \frac{d\mathbf{p}}{dt} = 0 + \mathbf{r} \times \mathbf{F} = \mathbf{r} \times \mathbf{F}$$

(The cross-product of velocity and momentum is zero, becausc these vectors are parallel.) So requiring the system to be "closed" here is mathematically equivalent to zero external torque acting on the system:

$$\mathbf{L}_{\text{system}} = \text{constant} \leftrightarrow \sum \tau_{\text{ext}} = 0$$

where τ_{ext} is any torque applied to the system of particles. It is assumed that internal interaction forces obey Newton's third law of motion in its strong form, that is, that the forces between particles are equal and opposite and act along the line between the particles.

In orbits, the angular momentum is distributed between the spin of the planet itself and the angular momentum of its orbit:

$$\mathbf{L}_{\text{total}} = \mathbf{L}_{\text{spin}} + \mathbf{L}_{\text{orbit}} \, ;$$

If a planet is found to rotate slower than expected, then astronomers suspect that the planet is accompanied by a satellite, because the total angular momentum is shared between the planet and its satellite in order to be conserved.

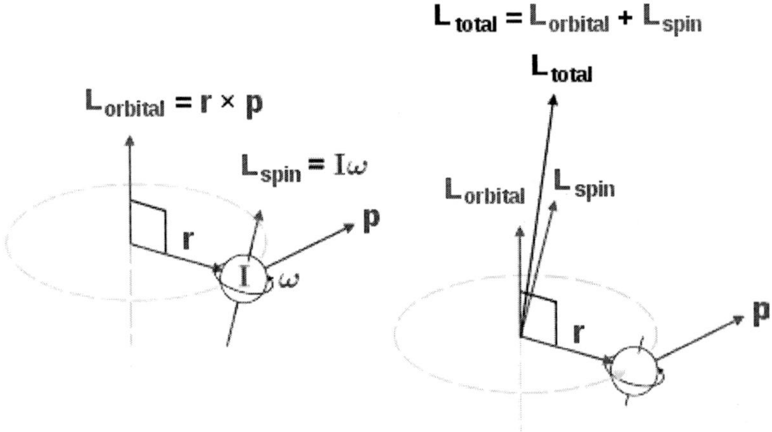

Conservation of angular momentum; distributed between the spin and orbital angular momenta. The moment of inertia and angular velocity of the spinning body are about its own spin axes, and its position is radial from the orbital axes and momentum tangential to the curve. Here circular orbits are shown - similar illustrations would follow for an elliptical orbit.

The conservation of angular momentum is used extensively in analysing what is called *central force motion*. If the net force on some body is directed always toward some fixed point, the *centre*, then there is no torque on the body with respect to the centre, and so the angular momentum of the body about the centre is constant. Constant angular momentum is extremely useful when dealing with the orbits of planets and satellites, and also when analysing the Bohr model of the atom.

The conservation of angular momentum explains the angular acceleration of an ice skater as she brings her arms and legs close to the vertical axis of rotation. By bringing part of mass of her body closer to the axis she decreases her body's moment of inertia. Because angular momentum is constant in the absence of external torques, the angular velocity (rotational speed) of the skater has to increase.

The same phenomenon results in extremely fast spin of compact stars (like white dwarfs, neutron stars and black holes) when they are formed out of much larger and slower rotating stars (indeed, decreasing the size of object 10^4 times results in increase of its angular velocity by the factor 10^8). The conservation of angular momentum in Earth–Moon system results in the transfer of angular momentum from Earth to Moon (due to tidal torque the Moon exerts on the Earth). This in turn results in the slowing down of the rotation rate of Earth (at about 42 nsec/day), and in gradual increase of the radius of Moon's orbit (at ~4.5 cm/year rate).

Angular Momentum in Relativistic Mechanics

In modern (late 20th century) theoretical physics, angular momentum is described using a different formalism. Under this formalism, angular momentum is the 2-form Noether charge associated with rotational invariance (As a result, angular momentum is not conserved for general curved spacetimes, unless it happens to be asymptotically rotationally invariant). For a system of point particles without any intrinsic angular momentum, it turns out to be

$$\sum_i \mathbf{r}_i \wedge \mathbf{p}_i$$

(Here, the wedge product is used.).

In the language of four-vectors and tensors the angular momentum of a particle in relativistic mechanics is expressed as an antisymmetric tensor of second order

$$L_{ij} = x_i \, p_j - x_j \, p_i$$

Angular Momentum in Quantum Mechanics

In quantum mechanics, angular momentum is quantized – that is, it cannot vary continuously, but only in "quantum leaps" between certain allowed values. The orbital angular momentum of a subatomic particle, that is due to its motion through space, is always a whole-number multiple of $\hbar$ ("h-bar," known as the reduced Planck's constant), which has an appropriate dimensionenergy time, or length momentum. Furthermore, experiments show that most subatomic particles have a permanent, built-in angular momentum, which is not due to their motion through space. This spin angular momentum comes in units of $\hbar/2$. For example, an electron standing at rest has an angular momentum of .

The classical definition of angular momentum as $\mathbf{L} = \mathbf{r} \times \mathbf{p}$ depends on six numbers: r_x, r_y, r_z, p_x, p_y, and p_z. Translating this into quantum-mechanical terms, the Heisenberg uncertainty principle tells us that it is not possible for all six of these numbers to be measured simultaneously with arbitrary precision. Therefore, there are limits to what can be known or measured about a particle's angular momentum. It turns out that the best that one can do is to simultaneously measure both the angular momentum vector's magnitude and its component along one axis. Mathematically, angular momentum in quantum mechanics is defined like momentum – not as a quantity but as an operator on the wave function:

$$\mathbf{L} = \mathbf{r} \times \mathbf{p}$$

where r and p are the position and momentum operators respectively. In particular, for a single particle with no electric charge and no spin, the angular momentum operator can be written in the position basis as

$$\mathbf{L} = -i\hbar(\mathbf{r} \times \nabla)$$

where ∇ is the vector differential operator del (also called "nabla"). This *orbital angular momentum operator* is the most commonly encountered form of the angular momentum operator, though not the only one. It satisfies the following canonical commutation relations:

$$[L_l, L_m] = i\hbar \sum_{n=1}^{3} \varepsilon_{lmn} L_n$$

where

- ε_{lmn} is the (antisymmetric) Levi-Civita symbol,
- $[X, Y] = XY$ " YX is the commutator.

From this follows

$$\left[L_i, L^2\right] = 0$$

Since,

$$L_x = -i\hbar\left(y\frac{\partial}{\partial z} - z\frac{\partial}{\partial y}\right)$$

$$L_y = -i\hbar\left(z\frac{\partial}{\partial x} - x\frac{\partial}{\partial z}\right)$$

$$L_z = -i\hbar\left(x\frac{\partial}{\partial y} - y\frac{\partial}{\partial x}\right)$$

it follows, for example,

$$\left[L_x, L_y\right] = -\hbar^2\left(\left(y\frac{\partial}{\partial z} - z\frac{\partial}{\partial y}\right)\left(z\frac{\partial}{\partial x} - x\frac{\partial}{\partial z}\right) - \left(z\frac{\partial}{\partial x} - x\frac{\partial}{\partial z}\right)\left(y\frac{\partial}{\partial z} - z\frac{\partial}{\partial y}\right)\right)$$

$$= -\hbar^2\left(y\frac{\partial}{\partial x} - x\frac{\partial}{\partial y}\right) = i\hbar L_z.$$

Addition of Quantized Angular Momenta

For more details on this topic, see Clebsch–Gordan coefficients.

Given a quantized total angular momentum $\vec{j}$ which is the sum of two individual quantized angular momenta $\vec{l_1}$ and $\vec{l_2}$,

$$\vec{j} = \vec{l_1} + \vec{l_2}$$

the quantum number j associated with its magnitude can range from | l_1 ” l_2 | to $l_1 + l_2$ in integer steps where l_1 and l_2 are quantum numbers corresponding to the magnitudes of the individual angular momenta.

Angular Momentum as a Generator of Rotations

If ϕ is the angle around a specific axis, for example the azimuthal angle around the z axis, then the angular momentum along this axis is the generator of rotations around this axis:

$$L_z = -i\hbar\frac{\partial}{\partial\phi}.$$

The eigenfunctions of L_z are therefore $e^{im_l\phi}$, and since ϕ has a period of 2π, m_l must be an integer.

For a particle with a spin S, this takes into account only the angular dependence of the location of the particle, for example its orbit in an atom. It is therefore known as orbital angular momentum. However, when one rotates the system, one also changes the spin. Therefore the total angular momentum, which is the full generator of rotations, is $J_i = L_i + S_i$ Being an angular momentum, J satisfies the same commutation relations as L, as will be explained below, namely

$$[J_\ell, J_m] = i\hbar\sum_{n=1}^{3}\varepsilon_{lmn}J_n$$

from which follows

$$\left[J_\ell, J^2\right] = 0.$$

Acting with J on the wavefunction ψ of a particle generates a rotation: $e^{i\phi J_z}\psi$ is the wavefunction ψ rotated around the z axis by an angle ϕ. For an infinitesimal rotation by an angle $d\phi$, the rotated wavefunction is $\psi + i d\phi J_z \psi$. This is similarly true for rotations around any axis.

In a charged particle the momentum gets a contribution from the electromagnetic field, and the angular momenta L and J change accordingly. If the Hamiltonian is invariant under rotations, as in spherically symmetric problems, then according to Noether's theorem, it commutes with the total angular momentum. So the total angular momentum is a conserved quantity

$$[J_l, H] = 0$$

Since angular momentum is the generator of rotations, its commutation relations follow the commutation relations of the generators of the three-dimensional rotation group SO(3). This is why J always satisfies these commutation relations. In *d dimensions*, the angular momentum will satisfy the same commutation relations as the generators of the *d*-dimensional rotation group SO(*d*). SO(3) has the same Lie algebra (i.e. the same commutation relations) as SU(2). Generators of SU(2) can have half-integer eigenvalues, and so can m_j. Indeed for fermions the spin S and total angular momentum J are half-integer. For *d*=3, in fact, j and m_j always are either integers or half-integers.

Technically, this is because the universal cover of SO(3) is isomorphic to SU(2), and the representations of the latter are fully known. J_i span the Lie algebra and J^2 is the Casimir invariant, and it can be shown that if the eigenvalues of J_z and J^2 are m_j and j(j+1) then m_j and j are both integer multiples of one-half. j is non-negative and m_j takes values between –j and j.

Relation to Spherical Harmonics

Angular momentum operators usually occur when solving a problem with spherical symmetry in spherical coordinates. Then, the angular momentum in space representation is:

$$L^2 = -\frac{\hbar^2}{\sin\theta}\frac{\partial}{\partial\theta}\left(\sin\theta\frac{\partial}{\partial\theta}\right) - \frac{\hbar^2}{\sin^2\theta}\frac{\partial^2}{\partial\phi^2}$$

When solving to find eigenstates of this operator, we obtain the following

$$L^2 \mid l,m\rangle = \hbar^2 l(l+1) \mid l,m\rangle$$

$$L_z \mid l,m\rangle - \hbar m \mid l,m\rangle$$

where

$$\langle \theta,\phi \mid l,m\rangle = Y_{l,m}(\theta,\phi)$$

are the spherical harmonics.

Thus, a particle whose wave function is the spherical harmonic $Y_{l,m}$ has an orbital angular momentum

$$L = \hbar\sqrt{l(l+1)}$$

with a z-component

$$L_z = \hbar m$$

Angular Momentum in Electrodynamics

When describing the motion of a charged particle in the presence of an electromagnetic field, the canonical momentum p is notgauge invariant. As a consequence, the canonical angular momentum $\mathbf{L} = \mathbf{r} \times \mathbf{p}$ is not gauge invariant either. Instead, the momentum that is physical, the so-called kinetic momentum, is

$$\mathbf{p} - \frac{e\mathbf{A}}{c}$$

where e is the electric charge, c the speed of light and A the vector potential. Thus, for example, the Hamiltonian of a charged particle of mass m in an electromagnetic field is then

$$H = \frac{1}{2m}\left(\mathbf{p} - \frac{e\mathbf{A}}{c}\right)^2 + e\phi$$

where ϕ is the scalar potential. This is the Hamiltonian that gives the Lorentz force law. The gauge-invariant angular momentum, or "kinetic angular momentum" is given by

$$K = \mathbf{r} \times \left(\mathbf{p} - \frac{e\mathbf{A}}{c}\right)$$

The interplay with quantum mechanics is discussed further in the article on canonical commutation relations.

Atomic Orbital

An atomic orbital is a mathematical function that describes the wave-like behaviour of either one electron or a pair of electrons in anatom. This function can be used to calculate the probability of finding any electron of an atom in any specific region around the atom's nucleus. The term may also refer to the physical region where the electron can be calculated to be, as defined by the particular mathematical form of the orbital.

Atomic orbitals are mathematical functions that depend on the coordinates of only *one* electron. They describe the wave-like nature of this electron in any energy state. They can be the hydrogen-like "orbitals" which are exact solutions to the Schrödinger equation for a hydrogen-like "atom" (i.e., the hydrogen atom or any ion formed by one electron and a nucleus). Alternatively, atomic orbitals refer to functions that depend on the coordinates of one electron (i.e. orbitals) but are used as starting points for approximating wave functions that depend on the simultaneous coordinates of all the electrons in an atom or molecule. The coordinate systems for atomic orbitals are usually spherical coordinates(r,θ,ö) in atoms and cartesians (x,y,z) in poly-atomic molecules.

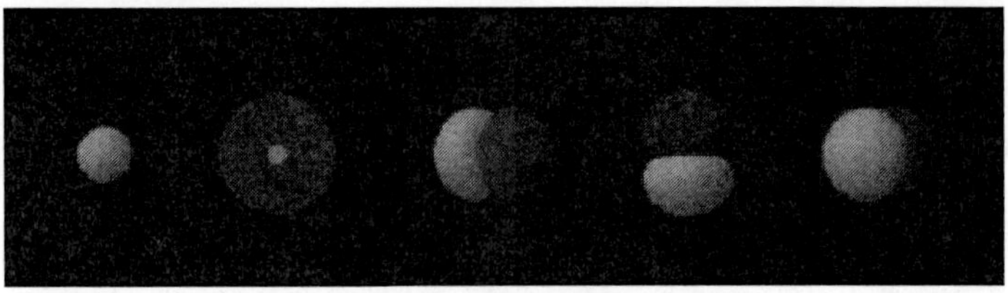

Figure: *The shapes of the first five atomic orbitals: 1s, 2s, $2p_x$, $2p_y$, and $2p_z$. The colours show the wave function phase. These are graphs of $\psi(x,y,z)$ functions which depend on the coordinates of one electron. To see the elongated shape of $\psi(x,y,z)^2$ functions that show probability density more directly.*

Within a visual context, atomic orbitals are the basic building blocks of the introductory pedagogical electron cloud model (derived from the wave mechanics model or atomic orbital model, but using particle concepts in order to visualize the mathematical procedures used to approximate wave functions for atoms with many electrons). Thus, (with particle concepts in *italics*) this model provides a framework for describing the *placement* of electrons in an atom. In this model, the atom consists of a nucleus surrounded by *orbiting* electrons. These electrons *exist in* so called atomic orbitals, which are a set of quantum states of the negatively charged electrons *trapped* in the electrical field generated by the positively-charged nucleus (which may be weakened by the effect of other electrons, but still remains attractive in sum). The electron cloud model can ultimately be described by quantum mechanics, in which the electrons are more accurately described as standing waves surrounding the nucleus.

Atomic orbitals are typically categorized by *n*, *l*, and *m* quantum numbers, which correspond to the electron's energy, angular momentum, and an angular momentum vector component, respectively. Each orbital is defined by a different set of quantum numbers and contains a maximum of two electrons. The simple names s orbital, p orbital, d orbital and f orbital refer to orbitals with angular momentum quantum number $l = 0$, 1, 2 and 3 respectively. These names indicate the orbital shape and are used to describe the electron configurations. They are derived from the characteristics of their spectroscopic lines: sharp, principal, diffuse, and fundamental, the rest being named in alphabetical order (omitting j).

The wave function for the *electron cloud* of a multi-electron atom may be seen as being built up (in approximation) in an electron configuration that is a product of simpler hydrogen-like atomic orbitals. $\varnothing_{He}(x_1,y_1,z_1,x_2,y_2,z_2)$ H" $1s(x_1,y_1,z_1) \cdot 1s(x_2,y_2,z_2) = 1s^2$. (This is read as, "The exact wave function depending on the simultaneous coordinates of the two electrons in the helium atom is approximated as a product of two one-s atomic orbitals each of which depends on the coordinates of only one electron.")

In such a configuration, pairs of electrons are arranged in simple repeating patterns of increasing odd numbers (1,3,5,7..), each of which represents a set of electron pairs with a given energy and angular momentum. The repeating *periodicity* of the blocks of 2, 6, 10, and 14 elements within sections of the periodic table arises naturally from the total number of electrons which occupy a complete set of s, p, d and f atomic orbitals, respectively.

There are typically three mathematical forms for atomic orbitals which can be chosen as a starting point for the calculation of the properties of atoms and molecules with many electrons. They all have forms that generate s, p, d, etc. functions.

Their differences are: 1) the *hydrogen-like* atomic orbitals are derived from the exact solution of the Schrödinger Equation for one electron and a nucleus. The part of the function that depends on the distance from the nucleus has nodes (radial nodes) and decays as $e^{(-distance)}$ from the nucleus. 2) The Slater-type orbital (STO) is a form without radial nodes but decays from the nucleus as does the hydrogen-like orbital. 3) The form of the Gaussian type orbital (Gaussians) has no radial nodes and decays as $e^{(-distance\ squared)}$. Although hydrogen like orbitals are still used as pedagogical tools, the advent of computers has made STOs preferable for atoms and diatomic molecules since combinations of STOs can replace the nodes in hydrogen-like atomic orbital. Gaussians are typically used in molecules with three or more atoms. Although not as accurate by themselves as STOs, combinations of many Gaussians can attain the accuracy of hydrogen-like orbitals.

With the development of quantum mechanics, it was found that the orbiting electrons around a nucleus could not be fully described as particles, but needed to be explained by the wave-particle duality. In this sense, the electrons have the following properties:

Wave-like Properties

The electrons do not orbit the nucleus in the sense of a planet orbiting the sun, but instead exist as standing waves. The lowest possible energy an electron can take is therefore analogous to the fundamental frequency of a wave on a string. Higher energy states are then similar to harmonics of the fundamental frequency.

The electrons are never in a single point location, although the probability of interacting with the electron at a single point can be found from the wave function of the electron.

Particle-like Properties

There is always an integer number of electrons orbiting the nucleus.

Electrons jump between orbitals in a particle-like fashion. For example, if a single photon strikes the electrons, only a single electron changes states in response to the photon.

The electrons retain particle like-properties such as: each wave state has the same electrical charge as the electron particle. Each wave state has a single discrete spin (spin up or spin down).

Visualizing Atomic Orbitals Intuitively

Despite the obvious analogy to planets revolving around the Sun, electrons cannot be described as solid particles. In addition, atomic orbitals do not closely resemble a planet's elliptical path in ordinary atoms. A more accurate analogy might be that of a large and often oddly-shaped "atmosphere" (the electron), distributed around a relatively tiny planet (the atomic nucleus). One difference is that some of an atom's electrons, those in s orbitals, have zero angular momentum, so they cannot in any sense be thought of as moving "around" the nucleus, as a planet does. (A planet would need to fall vertically into the Sun and oscillate up and down through it, to be in an orbit with no angular momentum). Other electrons do have varying amounts of angular momentum.

Atomic orbitals exactly describe the shape of this "atmosphere" only when a single electron is present in an atom. When more electrons are added to a single atom, the additional electrons tend to more evenly fill in a volume of space around the nucleus so that the resulting collection (sometimes termed the atom's "electron cloud") tends toward a generally spherical zone of probability describing where the atom's electrons will be found.

History

The term "orbital" was coined by Robert Mulliken in 1932. However, the idea that electrons might revolve around a compact nucleus with definite angular momentum was convincingly argued at least 19 years earlier by Niels Bohr, and the Japanese physicist Hantaro Nagaoka published an orbit-based hypothesis for electronic behaviour as early as 1904. Explaining the behaviour of these electron "orbits" was one of the driving forces behind the development of quantum mechanics.

Early Models

With J.J. Thomson's discovery of the electron in 1897, it became clear that atoms were not the smallest building blocks of nature, but were rather composite particles. The newly discovered structure within atoms tempted many to imagine how the atom's constituent parts might interact with each other. Thomson theorized that multiple electrons revolved in orbit-like rings within a positively-charged jelly-like substance, and between the electron's discovery and 1909, this "plum pudding model" was the most widely-accepted explanation of atomic structure.

Shortly after Thomson's discovery, Hantaro Nagaoka, a Japanese physicist, predicted a different model for electronic structure. Unlike the plum pudding model, the positive charge in Nagaoka's "Saturnian Model" was concentrated into a central core, pulling the electrons into circular orbits reminiscent of Saturn's rings. Few people took notice of Nagaoka's work at the time, and Nagaoka himself recognised a fundamental defect in the theory even at its conception, namely that a classical charged object cannot sustain orbital motion because it is accelerating and therefore loses energy due to electromagnetic radiation. Nevertheless, the Saturnian model turned out to have more in common with modern theory than any of its contemporaries.

The Bohr Atom

In 1909 Ernest Rutherford discovered that the positive half of atoms was tightly condensed into a nucleus, and it became clear from his analysis in 1911 that the plum pudding model could not explain atomic structure. Shortly after, in 1913, Rutherford's postdoctoral student Niels Bohr proposed a new model of the atom, wherein electrons orbited the nucleus with classical periods, but were only permitted to have discrete values of angular momentum, quantized in units $h/2\pi$. This constraint automatically permitted only certain values of electron energies. The Bohr model of the atom fixed the problem of energy loss from radiation from a ground state (by declaring that there was no state below this), and more importantly explained the origin of spectral lines.

After Bohr's use of Einstein's explanation of the photoelectric effect to relate energy levels in atoms with the wavelength of emitted light, the connection between the structure of electrons in atoms and the emission and absorption spectra of atoms became an increasingly useful tool in the understanding of electrons in atoms. The most prominent feature of emission and absorption spectra (known experimentally since the middle of the 19thcentury), was that these atomic spectra contained discrete lines. The significance of the Bohr model was that it related the lines in emission and absorption spectra to the energy differences between the orbits that electrons could take around an atom. This was, however, *not* achieved by Bohr through giving the electrons some kind of wave-like properties, since the idea that electrons could behave as matter waves was not suggested until twelve years later. Still, the Bohr model's use of quantized angular momenta and therefore quantized energy levels was a significant step towards the understanding of electrons in atoms, and also a significant step towards the development of quantum mechanics in suggesting that quantized restraints must account for all discontinuous energy levels and spectra in atoms.

With de Broglie's suggestion of the existence of electron matter waves in 1924, and for a short time before the full 1926 Schrödinger equation treatment of hydrogen like atom, a Bohr electron "wavelength" could be seen to be a function of its momentum, and thus a Bohr orbiting electron was seen to orbit in a circle at a multiple of its half-wavelength (this historically incorrect Bohr model is still occasionally taught to students). The Bohr model for a short time could be seen as a classical model with an additional constraint provided by the 'wavelength' argument. However, this period was immediately superseded by the full three-dimentional wave mechanics of 1926. In our current understanding of physics, the Bohr model is called a semi-classical model because of its quantization of angular momentum, not primarily because of its relationship with electron wavelength, which appeared in hindsight a dozen years after the Bohr model was proposed.

The Bohr model was able to explain the emission and absorption spectra of hydrogen. The energies of electrons in the n=1, 2, 3, etc. states in the Bohr model match those of current physics. However, this did not explain similarities between different atoms, as expressed by the periodic table, such as the fact that helium (2 electrons), neon (10 electrons), and argon (18 electrons) exhibit similar chemical behaviour. Modern physics explains this by noting that the n=1 state holds 2 electrons, the n=2 state holds 8 electrons, and the n=3 state holds 8 electrons (in argon). In the end, this was solved by the discovery of modern quantum mechanics and the Pauli Exclusion Principle.

Modern Conceptions and Connections to the Heisenberg Uncertainty Principle

Immediately after Heisenberg discovered his uncertainty relation, it was noted by Bohr that the existence of any sort of wave packet implies uncertainty in the wave frequency and wavelength, since a spread of frequencies is needed to create the packet itself. In quantum mechanics, where all particle momenta are associated with waves, it is the formation of such a wave packet which localises the wave, and thus the particle, in space. In states where a quantum mechanical particle is bound, it must be localised as a wave packet, and the existence of the packet and its minimum size implies a spread and minimal value in particle wavelength, and thus also momentum and energy. In quantum mechanics, as a particle is localised to a smaller region in space, the associated compressed wave packet requires a larger and larger range of momenta, and thus larger kinetic energy. Thus, the binding energy to contain or trap a particle in a smaller region of space, increases without bound, as the region of space grows smaller. Particles cannot be restricted to a geometric point in space, since this would require an infinite particle momentum.

In chemistry, Schrödinger, Pauling, Mulliken and others noted that the consequence of Heisenberg's relation was that the electron, as a wave packet, could not be considered to have an exact location in its orbital. Max Born suggested that the electron's position needed to be described by a probability distribution which was connected with finding the electron at some point in the wave-function which described its associated wave packet. The new quantum mechanics did not give exact results, but only the probabilities for the occurrence of a variety of possible such results. Heisenberg held that the path of a moving particle has no meaning if we cannot observe it, as we cannot with electrons in an atom.

In the quantum picture of Heisenberg, Schrödinger and others, the Bohr atom number n for each orbital became known as an *n-sphere* in a three dimensional atom and was pictured as the mean energy of the probability cloud of the electron's wave packet which surrounded the atom.

Orbital Names

Orbitals are given names in the form:

$$X\text{type}^y$$

where X is the energy level corresponding to the principal quantum number n, type is a lower-case letter denoting the shape or subshell of the orbital and it corresponds to the angular quantum number l, and y is the number of electrons in that orbital.

For example, the orbital $1s^2$ (pronounced "one ess two") has two electrons and is the lowest energy level ($n = 1$) and has an angular quantum number of $l = 0$. In X-ray notation, the *principal quantum number* is given a letter associated with it. For $n = 1, 2, 3, 4, 5, ...,$ the letters associated with those numbers are $K, L, M, N, O, ...,$ respectively.

Formal Quantum Mechanical Definition

In quantum mechanics, the state of an atom, i.e. the eigenstates of the atomic Hamiltonian, is expanded into linear combinations of anti-symmetrized products (Slater determinants) of one-electron functions. The spatial components of these one-electron functions are called atomic orbitals. (When one considers also their spin component, one speaks of atomic spin orbitals.)

In atomic physics, the atomic spectral lines correspond to transitions (quantum leaps) between quantum states of an atom. These states are labelled by a set of quantum numbers summarised in the term symbol and usually associated with particular electron configurations, i.e., by occupation schemes of atomic orbitals (e.g., $1s^2\ 2s^2\ 2p^6$ for the ground state of neon — term symbol: $^1S_0$).

This notation means that the corresponding Slater determinants have a clear higher weight in the configuration interaction expansion. The atomic orbital concept is therefore a key concept for visualizing the excitation process associated with a given transition. For example, one can say for a given transition that it corresponds to the excitation of an electron from an occupied orbital to a given unoccupied orbital. Nevertheless, one has to keep in mind that electrons are fermions ruled by the Pauli exclusion principle and cannot be distinguished from the other electrons in the atom. Moreover, it sometimes happens that the configuration interaction expansion converges very slowly and that one cannot speak about simple one-determinant wave function at all. This is the case when electron correlation is large. Fundamentally, an atomic orbital is a one-electron wave function, even though most electrons do not exist in one-electron atoms, and so the one-electron view is an approximation. When thinking about orbitals, we are often given an orbital vision which (even if it is not spelled out) is heavily influenced by this Hartree–Fock approximation, which is one way to reduce the complexities of molecular orbital theory.

Hydrogen-like Atoms

The simplest atomic orbitals are those that are calculated for systems with a single electron, such as the hydrogen atom. An atom of any other element ionized down to a single electron is very similar to hydrogen, and the orbitals take the same form. In the Schrödinger equation for this system of one negative and one positive particle, the atomic orbitals are the eigenstates of the Hamiltonian operator for the energy. They can be obtained analytically, meaning that the resulting orbitals are products of a polynomial series, and exponential and trigonometric functions. For atoms with two or more electrons, the governing equations can only be solved with the use of methods of iterative approximation. Orbitals of multi-electron atoms are *qualitatively* similar to those of hydrogen, and in the simplest models, they are taken to have the same form. For more rigorous and precise analysis, the numerical approximations must be used.

A given (hydrogen-like) atomic orbital is identified by unique values of three quantum numbers: n, l, and m_l. The rules restricting the values of the quantum numbers, and their energies, explain the electron configuration of the atoms and the periodic table. The stationary states (quantum states) of the hydrogen-like atoms are its atomic orbitals. However, in general, an electron's behaviour is not fully described by a single orbital. Electron states are best represented by time-depending "mixtures" (linear combinations) of multiple orbitals. See Linear combination of atomic orbitals molecular orbital method.

The quantum number n first appeared in the Bohr model where it determines the radius of each circular electron orbit. In modern quantum mechanics however, n determines the mean distance of the electron from the nucleus; all electrons with the same value of n lie at the same average distance. For this reason, orbitals with the same value of n are said to comprise a "shell". Orbitals with the same value of n and also the same value of l are even more closely related, and are said to comprise a "subshell".

Quantum Numbers

Because of the quantum mechanical nature of the electrons around a nucleus, they cannot be described by a location and momentum. Instead, they are described by a set of quantum numbers that encompasses both the particle-like nature and the wave-like nature of the electrons. An atomic orbital is uniquely identified by the values of the three quantum numbers, and each set of the three quantum numbers corresponds to exactly one orbital, but the quantum numbers only occur in certain combinations of values. The quantum numbers, together with the rules governing their possible values, are as follows:

The principal quantum number, n, describes the energy of the electron and is always a positive integer. In fact, it can be any positive integer, but for reasons discussed below, large numbers are seldom encountered. Each atom has, in general, many orbitals associated with each value of n; these orbitals together are sometimes called *electron shells*. The azimuthal quantum number, ℓ, describes the orbital angular momentum of each electron and is a non-negative integer. Within a shell where n is some integer n_0, ℓ ranges across all (integer) values satisfying the relation $0 \leq \ell \leq n_0 - 1$. For instance, the n = 1 shell has only orbitals with $\ell = 0$, and the n = 2 shell has only orbitals with , and . The set of orbitals associated with a particular value of re sometimes collectively called a *subshell.*

The magnetic quantum number, m_ℓ, describes the magnetic moment of an electron in an arbitrary direction, and is also always an integer. Within a subshell where ℓ is some integer ℓ_0, m_ℓ ranges thus: $-\ell_0 \leq m_\ell \leq \ell_0$.

The above results may be summarised in the following table. Each cell represents a subshell, and lists the values of m_ℓ available in that subshell. Empty cells represent subshells that do not exist.

	l=0	l=1	l=2	l=3	l=4	...
n=1	m_l = 0					
n=2	0	-1, 0, 1				
n=3	0	-1, 0, 1	-2, -1, 0, 1, 2			
n=4	0	-1, 0, 1	-2, -1, 0, 1, 2	-3, -2, -1, 0, 1, 2, 3		
n=5	0	-1, 0, 1	-2, -1, 0, 1, 2	-3, -2, -1, 0, 1, 2, 3	-4, -3, -2 -1, 0, 1, 2, 3, 4	
...	...	...	...	...	...	...

Subshells are usually identified by their n- and ℓ-values. n is represented by its numerical value, but ℓ is represented by a letter as follows: 0 is represented by 's', 1 by 'p', 2 by 'd', 3 by 'f', and 4 by 'g'. For instance, one may speak of the subshell with n = 2 and $\ell = 0$ as a '2s subshell'.

Each electron also has a spin quantum number, s, which describes the spin of each electron (spin up or spin down). The number s can be +^{1}D$_{2}$ or -^{1}D$_{2}$.

The Pauli exclusion principle states that no two electrons can occupy the same quantum state: every electron in an atom must have a unique combination of quantum numbers.

The Shapes of Orbitals

Cross-section of computed hydrogen atom orbital ($\psi(r,\theta,\ddot{o})^2$) for the 6s ($n$=6, l=0,m=0) orbital. Note that s orbitals, though spherically symmetrical, have radially placed wave-nodes for $n > 1$. However, only s orbitals invariably have a centre anti-node; the other types never do.

Any discussion of the shapes of electron orbitals is necessarily arbitrary since a number representing a certain probability of finding the electron must be chosen (usually from zero up to one, or from 0 up to 100%). According to the uncertainty principle and the hypotheses of Quantum Mechanics a given electron has one and only one probability of being found at any moment at a particular point (x,y,z) in space. For an atom, it is easier to calculate the probability using a distance r from the nucleus and a direction (θ,φ) using polar coordinates. If an atomic orbital is used to describe this electron, then the probability of finding an electron at point (r,θ,φ) is obtained from this orbital, which is the mathematical function $\psi(r,\theta,\varphi)$. However, the electron is much more likely to be found in particular regions around the atom than in others. Given this, a boundary surface can be drawn so that the electron has a high probability to be found somewhere within the surface, and all regions outside the surface have low values. Even though the precise placement of the surface is arbitrary, any reasonably compact determination must follow a pattern specified by the behaviour of $|\psi(r,\theta,\varphi)|^2$, the square of the absolute value (also called magnitude or modulus) of the complex-valued wave function. This boundary surface is sometimes what is meant when the "shape" of an orbital is referred to. Sometimes the ψ function will be graphed to show its phases, rather than the $|\psi(r,\theta,\varphi)|^2$ which shows probability density but has no phases (which have been lost in the process of taking the absolute value, since $\psi(r,\theta,\varphi)$ is a complex number). $|\psi(r,\theta,\varphi)|^2$ orbital graphs tend to have less spherical, thinner lobes than $\psi(r,\theta,\varphi)$ graphs, but have the same number of lobes in the same places, and otherwise are recognisable. This article, in order to show wave function phases, shows mostly y(r,q,j) graphs.

The lobes can be viewed as interference patterns between the two counter rotating "*m*" and "-*m*" modes, with the projection of the orbital onto the xy plane having a resonant "*m*" wavelengths around the circumference. For each *m* there are two of these <*m*>+<-*m*> and <*m*>-<-*m*>. For the case where m=0 the orbital is vertical, counter rotating information is unknown, and the orbital is z-axis symmetric. For the case where *l=0* there are no counter rotating modes. There are only radial modes and the shape is spherically symmetric. For any given *n*, the smaller *l* is, the more radial nodes there are. Loosely speaking *n* is energy, *l* is analogous to eccentricity, and *m* is orientation. Generally speaking, the number *n* determines the size and energy of the orbital for a given nucleus: as n increases, the size of the orbital increases. However, in comparing different elements, the higher nuclear charge, *Z*, of heavier elements causes their orbitals to contract by comparison to lighter ones, so that the overall size of the whole atom remains very roughly constant, even as the number of electrons in heavier elements (higher *Z*) increases. Also in general terms, ℓ determines an orbital's shape, and m_ℓ its orientation. However, since some orbitals are described by equations in complex numbers, the shape sometimes depends on m_ℓ also.

The single s-orbitals $\ell = 0$ are shaped like spheres. For $n = 1$ the sphere is "solid" (it is most dense at the centre and fades exponentially outwardly), but for $n = 2$ or more, each single s-orbital is composed of spherically symmetric surfaces which are nested shells (i.e., the "wave-structure" is radial, following a sinusoidal radial component as well). The s-orbitals for all n numbers are the only orbitals with an anti-node (a region of high wave function density) at the centre of the nucleus. All other orbitals (p, d, f, etc.) have angular momentum, and thus avoid the nucleus (having a wave node *at* the nucleus).

The three p-orbitals for $n = 2$ have the form of two ellipsoids with a point of tangency at the nucleus (the two-lobed shape is sometimes referred to as a "dumbbell"). The three p-orbitals in each shell are oriented at right angles to each other, as determined by their respective linear combination of values of m_ℓ.

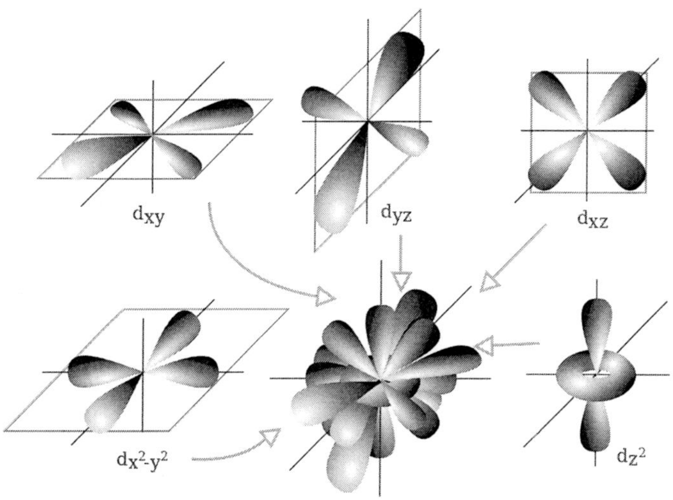

Figure: *The five d orbitals in $\psi(x,y,z)^2$ form, with a combination diagram showing how they fit together to fill space around an atomic nucleus.*

Four of the five d-orbitals for $n = 3$ look similar, each with four pear-shaped lobes, each lobe tangent to two others, and the centres of all four lying in one plane, between a pair of axes. Three of these planes are the xy-, xz-, and yz-planes, and the fourth has the centres on the x and y axes. The fifth and final d-orbital consists of three regions of high probability density: a torus with two pear-shaped regions placed symmetrically on its z axis.

There are seven f-orbitals, each with shapes more complex than those of the d-orbitals.

For each s, p, d, f and g set of orbitals, the set of orbitals which composes it forms a spherically symmetrical set of shapes. For non-s orbitals, which have lobes, the lobes point in directions so as to fill space as symmetrically as possible for number of lobes which exist for a set of orientations.

For example, the three p orbitals have six lobes which are oriented to each of the six primary directions of 3-D space; for the 5 d orbitals, there are a total of 18 lobes, in which again six point in primary directions, and the 12 additional lobes fill the 12 gaps which exist between each pairs of these 6 primary axes.

Additionally, as is the case with the s orbitals, individual p, d, f and g orbitals with n values higher than the lowest possible value, exhibit an additional radial node structure which is reminiscent of harmonic waves of the same type, as compared with the lowest (or fundamental) mode of the wave.

As with s orbitals, this phenomenon provides p, d, f, and g orbitals at the next higher possible value of n (for example, 3p orbitals vs. the fundamental 2p), an additional node in each lobe. Still higher values of n further increase the number of radial nodes, for each type of orbital.

The shapes of atomic orbitals in one-electron atom are related to 3-dimensional spherical harmonics. These shapes are not unique, and any linear combination is valid, like a transformation to cubic harmonics, in fact it is possible to generate sets where all the d's are the same shape, just like the p_x,p_y, and p_z are the same shape.

Understanding Why Atomic Orbitals Take These Shapes

The shapes of atomic orbitals can be understood qualitatively by considering the analogous case of standing waves on a circular drum. To see the analogy, the mean vibrational displacement of each bit of drum membrane from the equilibrium point over many cycles (a measure of average drum membrane velocity and momentum at that point) must be considered relative to that point's distance from the centre of the drum head. If this displacement is taken as being analogous to the probability of finding an electron at a given distance from the nucleus, then it will be seen that the many modes of the vibrating disk form patterns that trace the various shapes of atomic orbitals. The basic reason for this correspondence lies in the fact that the distribution of kinetic energy and momentum in a matter-wave is predictive of where the particle associated with the wave will be. That is, the probability of finding an electron at a given place is also a function of the electron's average momentum at that point, since high electron momentum at a given position tends to "localise" the electron in that position, via the properties of electron wave-packets.

This relationship means that certain key features can be observed in both drum membrane modes and atomic orbitals. For example, in all of the modes analogous to s orbitals (the top row in the illustration), it can be seen that the very centre of the drum membrane vibrates most strongly, corresponding to the antinode in all s orbitals in an atom. This antinode means the electron is most likely to be at the physical position of the nucleus (which it passes straight through without scattering or striking it), since it is moving (on average) most rapidly at that point, giving it maximal momentum.

A mental "planetary orbit" picture closest to the behaviour of electrons in s orbitals, all of which have no angular momentum, might perhaps be that of the path of an atomic-sized black hole, or some other imaginary particle which is able to fall with increasing velocity from space directly through the Earth, without stopping or being affected by any force but gravity, and in this way falls through the core and out the other side in a straight line, and off again into space, while slowing from the backwards gravitational tug.

If such a particle were gravitationally bound to the Earth it would not escape, but would pursue a series of passes in which it always slowed at some maximal distance into space, but had its maximal velocity at the Earth's centre (this "orbit" would have an orbital eccentricity of 1.0). If such a particle also had a wave nature, it would have the highest probability of being located where its velocity and momentum were highest, which would be at the Earth's core.

In addition, rather than be confined to an infinitely narrow "orbit" which is a straight line, it would pass through the Earth from all directions, and not have a preferred one. Thus, a "long exposure" photograph of its motion over a very long period of time, would show a sphere.

In order to be stopped, such a particle would need to interact with the Earth in some way other than gravity. In a similar way, all s electrons have a finite probability of being found inside the nucleus, and this allows s electrons to occasionally participate in strictly nuclear-electron interaction processes, such as electron capture and internal conversion.

Orbital Energy

In atoms with a single electron (hydrogen-like atoms), the energy of an orbital (and, consequently, of any electrons in the orbital) is determined exclusively by n. The $n = 1$ orbital has the lowest possible energy in the atom. Each successively higher value of n has a higher level of energy, but the difference decreases as n increases. For high n, the level of energy becomes so high that the electron can easily escape from the atom. In single electron atoms, all levels with different ℓ within a given n are (to a good approximation) degenerate, and have the same energy. In atoms with multiple electrons, the energy of an electron depends not only on the intrinsic properties of its orbital, but also on its interactions with the other electrons.

These interactions depend on the detail of its spatial probability distribution, and so the energy levels of orbitals depend not only on n but also on ℓ. Higher values of are associated with higher values of energy; for instance, the 2p state is higher than the 2s state. When = 2, the increase in energy of the orbital becomes so large as to push the energy of orbital above the energy of the s-orbital in the next higher shell; when = 3 the energy is pushed into the shell two steps higher. The filling of the 3d orbitals does not occur until the 4s orbitals have been filled.

The increase in energy for subshells of increasing angular momentum in larger atoms is due to electron-electron interaction effects, and it is specifically related to the ability of low angular momentum electrons to penetrate more effectively toward the nucleus, where they are subject to less screening from the charge of intervening electrons. Thus, in atoms of higher atomic number, the ℓ of electrons becomes more and more of a determining factor in their energy, and the principal quantum numbers n of electrons becomes less and less important in their energy placement.

The energy sequence of the first 24 subshells (e.g., 1s, 2p, 3d, etc.) is given in the following table. Each cell represents a subshell with n and given by its row and column indices, respectively. The number in the cell is the subshell's position in the sequence. For a linear listing of the subshells in terms of increasing energies in multielectron atoms.

	s	*p*	*d*	*f*	*g*
1	1				
2	2	3			
3	4	5	7		
4	6	8	10	13	
5	9	11	14	17	*21*
6	12	15	18	*22*	
7	16	19	*23*		
8	*20*	*24*			

Note: empty cells indicate non-existent sublevels, while numbers in italics indicate sublevels that could exist, but which do not hold electrons in any element currently known.

Electron Placement and the Periodic Table

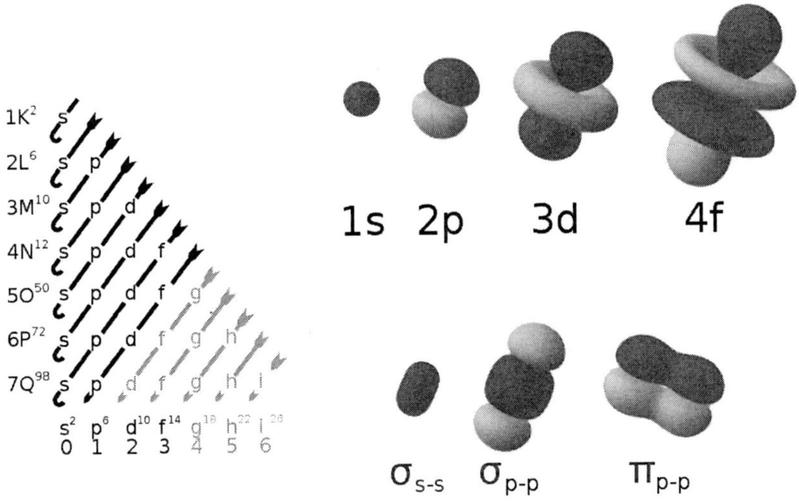

$1s^2_2 2s^2_4 2p^6_{10} 3s^2_{12} 3p^6_{18} 4s^2_{20} 3d^{10}_{30} 4p^6_{36} 5s^2_{38} 4d^{10}_{48} 5p^6_{54} 6s^2_{56} 4f^{14}_{70} 5d^{10}_{80} 6p^6_{86} 7s^2_{88} 5f^{14}_{102} 6d^{10}_{112} 7p^6_{118}$

***Figure:** Electron atomic and molecular orbitals. The chart of orbitals (left) is arranged by increasing energy (Madelung rule). Note that atomic orbits are functions of three variables (two angles, and the distance from the nucleus, r). These images are faithful to the angular component of the orbital, but not entirely representative of the orbital as a whole.*

Several rules govern the placement of electrons in orbitals (*electron configuration*). The first dictates that no two electrons in an atom may have the same set of values of quantum numbers (this is the Pauli exclusion principle). These quantum numbers include the three that define orbitals, as well as *s*, or spin quantum number. Thus, two electrons may occupy a single orbital, so long as they have different values of *s*. However, *only* two electrons, because of their spin, can be associated with each orbital.

Additionally, an electron always tends to fall to the lowest possible energy state. It is possible for it to occupy any orbital so long as it does not violate the Pauli exclusion principle, but if lower-energy orbitals are available, this condition is unstable. The electron will eventually lose energy (by releasing a photon) and drop into the lower orbital. Thus, electrons fill orbitals in the order specified by the energy sequence given above.

This behaviour is responsible for the structure of the periodic table. The table may be divided into several rows (called 'periods'), numbered starting with 1 at the top. The presently known elements occupy seven periods. If a certain period has number i, it consists of elements whose outermost electrons fall in the ith shell. Niels Bohr was the first to propose (1923) that the periodicity in the properties of the elements might be explained by the periodic filling of the electron energy levels, resulting in the electronic structure of the atom.

The periodic table may also be divided into several numbered rectangular 'blocks'. The elements belonging to a given block have this common feature: their highest-energy electrons all belong to the same ℓ-state (but the n associated with that -state depends upon the period). For instance, the leftmost two columns constitute the 's-block'. The outermost electrons of Li and Be respectively belong to the 2s subshell, and those of Na and Mg to the 3s subshell.

The following is the order for filling the "subshell" orbitals, which also gives the order of the "blocks" in the periodic table:

1s, 2s, 2p, 3s, 3p, 4s, 3d, 4p, 5s, 4d, 5p, 6s, 4f, 5d, 6p, 7s, 5f, 6d, 7p

The "periodic" nature of the filling of orbitals, as well as emergence of the s, p, d and f "blocks" is more obvious, if this order of filling is given in matrix form, with increasing principal quantum numbers starting the new rows ("periods") in the matrix. Then, each subshell (composed of the first two quantum numbers) is repeated as many times as required for each pair of electrons it may contain. The result is a compressed periodic table, with each entry representing two successive elements:

1s									
2s							2p	2p	2p
3s							3p	3p	3p
4s		3d	3d	3d	3d	3d	4p	4p	4p
5s		4d	4d	4d	4d	4d	5p	5p	5p
6s	(4f)	5d	5d	5d	5d	5d	6p	6p	6p
7s	(5f)	6d	6d	6d	6d	6d	7p	7p	7p

The number of electrons in an electrically neutral atom increases with the atomic number. The electrons in the outermost shell, or *valence electrons*, tend to be responsible for an element's chemical behaviour. Elements that contain the same number of valence electrons can be grouped together and display similar chemical properties.

Relativistic Effects

For elements with high atomic number Z, the effects of relativity become more pronounced, and especially so for *s* electrons, which move at relativistic velocities as they penetrate the screening electrons near the core of high Z atoms. This relativistic increase in momentum for high speed electrons causes a corresponding decrease in wavelength and contraction of 6s orbitals relative to 5d orbitals (by comparison to corresponding *s* and *d* electrons in lighter elements in the same column of the periodic table); this results in 6s valence electrons becoming lowered in energy.

Examples of significant physical outcomes of this effect include the lowered melting temperature of mercury (which results from 6s electrons not being available for metal bonding) and the golden colour of gold and caesium (which results from narrowing of 6s to 5d transition energy to the point that visible light begins to be absorbed).

In the Bohr Model, an $n = 1$ electron has a velocity given by $v = Z\alpha c$, where Z is the atomic number, α is the fine-structure constant, and c is the speed of light. In non-relativistic quantum mechanics, therefore, any atom with an atomic number greater than 137 would require its 1s electrons to be travelling faster than the speed of light.

Even in the Dirac equation, which accounts for relativistic effects, the wavefunction of the electron for atoms with $Z > 137$ is oscillatory and unbounded. The significance of element 137, also known as untriseptium, was first pointed out by the physicist Richard Feynman. Element 137 is sometimes informally called feynmanium (symbol Fy). However, Feynman's approximation fails to predict the exact critical value of Z due to the non-point-charge nature of the nucleus and very small orbital radius of inner electrons, resulting in a potential seen by inner electrons which is effectively less than Z. The critical Z value which makes the atom unstable with regard to high-field breakdown of the vacuum and production of electron-positron pairs, does not occur until Z is about 173. These conditions are not seen except transiently in collisions of very heavy nuclei such as lead or uranium in accelerators, where such electron-positron production from these effects has been claimed to be observed. See Extension of the periodic table beyond the seventh period.

There are no nodes in relativistic orbital densities, although individual components of the wavefunction will have nodes.

Transitions between Orbitals

Under quantum mechanics, each quantum state has a well-defined energy. When applied to atomic orbitals, this means that each state has a specific energy, and that if an electron is to move between states, the energy difference is also very fixed.

Consider two states of the Hydrogen atom:

State 1) n=1, l=0, m_l=0 and s=+[1]D $_2$

State 2) n=2, l=0, m_l=0 and s=+[1]D $_2$

By quantum theory, state 1 has a fixed energy of E_1, and state 2 has a fixed energy of E_2. Now, what would happen if an electron in state 1 were to move to state 2? For this to happen, the electron would need to gain an energy of exactly E_2 - E_1. If the electron receives energy that is less than or greater than this value, it cannot jump from state 1 to state 2. Now, suppose we irradiate the atom with a broad-spectrum of light. Photons that reach the atom that have an energy of exactly E_2 - E_1 will be absorbed by the electron in state 1, and that electron will jump to state 2. However, photons that are greater or lower in energy cannot be absorbed by the electron, because the electron can only jump to one of the orbitals, it cannot jump to a state between orbitals. The result is that only photons of a specific frequency will be absorbed by the atom. This creates a line in the spectrum, known as an absorption line, which corresponds to the energy difference between states 1 and 2.

The atomic orbital model thus predicts line spectra, which are observed experimentally. This is one of the main validations of the atomic orbital model.

The atomic orbital model is nevertheless an approximation to the full quantum theory, which only recognises many electron states. The predictions of line spectra are qualitatively useful but are not quantitatively accurate for atoms and ions other than those containing only one electron.

Wave–Particle Duality

Wave–particle duality postulates that all particles exhibit bothwave and particle properties. A central concept of quantum mechanics, this duality addresses the inability of classical concepts like “particle” and “wave” to fully describe the behaviour of quantum-scale objects. Standard interpretations of quantum mechanics explain this paradox as a fundamental property of the Universe, while alternative interpretations explain the duality as an emergent, second-order consequence of various limitations of the observer. This treatment focuses on explaining the behaviour from the perspective of the widely used Copenhagen interpretation, in which wave–particle duality is one aspect of the concept of complementarity, that a phenomenon can be viewed in one way or in another, but not both simultaneously.

The idea of duality originated in a debate over the nature of light and matter that dates back to the 17th century, when competing theories of light were proposed by Christiaan Huygens and Isaac Newton: light was thought either to consist of waves (Huygens) or of particles (Newton).

Through the work of Max Planck, Albert Einstein, Louis de Broglie, Arthur Compton, Niels Bohr, and many others, current scientific theory holds that *all* particles also have a wave nature (and vice versa). This phenomenon has been verified not only for elementary particles, but also for compound particles like atoms and even molecules. In fact, according to traditional formulations of non-relativistic quantum mechanics, wave–particle duality applies to all objects, even macroscopicones; but because of their small wavelengths, the wave properties of macroscopic objects cannot be detected.

Brief History of Wave and Particle Viewpoints

Aristotle was one of the first to publicly hypothesize about the nature of light, proposing that light is a disturbance in the element air. (That is, it is a wave-like phenomenon). On the other hand, Democritus– the original *atomist* – argued that all things in the universe, including light, are composed of indivisible sub-components (light being some form of solar atom). At the beginning of the 11th Century, the Arabic scientist Alhazen wrote the first comprehensive treatise on optics; describing refraction, reflection, and the operation of a pinhole lens via rays of light travelling from the point of emission to the eye. He asserted that these rays were composed of particles of light. In 1630, René Descartes popularized and accredited in the West the opposing wave description in *his* treatise on light, showing that the behaviour of light could be re-created by modelling wave-like disturbances in his universal medium ("plenum"). Beginning in 1670 and progressing over three decades, Isaac Newton developed and championed *his* corpuscular hypothesis, arguing that the perfectly straight lines of reflection demonstrated light's particle nature; only particles could travel in such straight lines. He explained refraction by positing that particles of light accelerated laterally upon entering a denser medium. Around the same time, Newton's contemporaries Robert Hooke and Christian Huygens – and later Augustin-Jean Fresnel – mathematically refined the wave viewpoint, showing that if light travelled at different speeds in different media (such as water and air), refraction could be easily explained as the medium-dependent propagation of light waves. The resulting Huygens–Fresnel principle was extremely successful at reproducing light's behaviour and, subsequently supported by Thomas Young's discovery of double-slit interference, was the beginning of the end for the particle light camp.

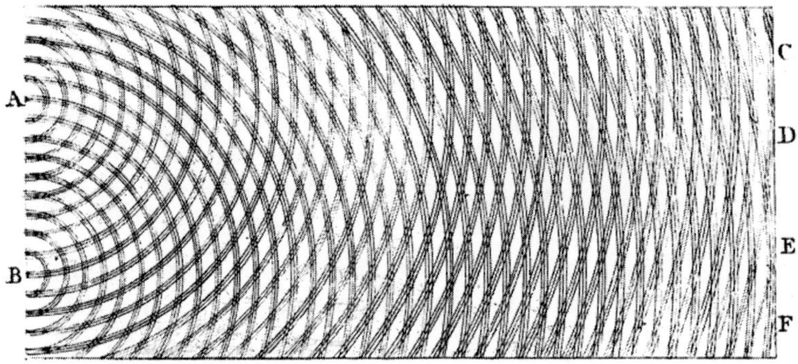

Figure: *Thomas Young's sketch of two-slit diffraction of waves, 1803.*

The final blow against corpuscular theory came when James Clerk Maxwell discovered that he could combine four simple equations, which had been previously discovered, along with a slight modification to describe self propagating waves of oscillating electric and magnetic fields. When the propagation speed of these electromagnetic waves was calculated, the speed of light fell out. It quickly became apparent that visible light, ultraviolet light, and infrared light (phenomena thought previously to be unrelated) were all electromagnetic waves of differing frequency. The wave theory had prevailed – or at least it seemed to.

While the 19th century had seen the success of the wave theory at describing light, it had also witnessed the rise of the atomic theory at describing matter. In 1789, Antoine Lavoisier securely differentiated chemistry from alchemy by introducing rigor and precision into his laboratory techniques; allowing him to deduce the conservation of mass and categorize many new chemical elements and compounds. However, the nature of these essential chemical elements remained unknown. In 1799, Joseph Louis Proust advanced chemistry towards the atom by showing that elements combined indefinite proportions.

This led John Dalton to resurrect Democritus' atom in 1803, when he proposed that elements were invisible sub components; which explained why the varying oxides of metals (e.g.stannous oxide and cassiterite, SnO and SnO_2 respectively) possess a 1:2 ratio of oxygen to one another. But Dalton and other chemists of the time had not considered that some elements occur in monatomic form (like Helium) and others in diatomic form (like Hydrogen), or that water was H_2O, not the simpler and more intuitive HO – thus the atomic weights presented at the time were varied and often incorrect. Additionally, the formation of HO by two parts of hydrogen gas and one part of oxygen gas would require an atom of oxygen to split in half (or two half-atoms of hydrogen to come together).

This problem was solved by Amedeo Avogadro, who studied the reacting volumes of gases as they formed liquids and solids. By postulating that equal volumes of elemental gas contain an equal number of atoms, he was able to show that H_2O was formed from two parts H_2 and one part O_2.

By discovering diatomic gases, Avogadro completed the basic atomic theory, allowing the correct molecular formulae of most known compounds – as well as the correct weights of atoms – to be deduced and categorized in a consistent manner. The final stroke in classical atomic theory came when Dimitri Mendeleev saw an order in recurring chemical properties, and created a table presenting the elements in unprecedented order and symmetry. But there were holes in Mendeleev's table, with no element to fill them in. His critics initially cited this as a fatal flaw, but were silenced when new elements were discovered that perfectly fit into these holes. The success of the periodic table effectively converted any remaining opposition to atomic theory; even though no single atom had ever been observed in the laboratory, chemistry was now an atomic science.

The Turn of the 19th Century and the Paradigm Shift

Particles of Electricity?

At the close of the 19th century, the reductionism of atomic theory began to advance into the atom itself; determining, through physics, the nature of the atom and the operation of chemical reactions.

Electricity, first thought to be a fluid, was now understood to consist of particles called electrons. This was first demonstrated by J. J. Thomson in 1897 when, using a cathode ray tube, he found that an electrical charge would travel across a vacuum (which would possess infinite resistance in classical theory).

Since the vacuum offered no medium for an electric fluid to travel, this discovery could only be explained via a particle carrying a negative charge and moving through the vacuum. This *electron* flew in the face of classical electrodynamics, which had successfully treated electricity as a fluid for many years (leading to the invention of batteries, electric motors, dynamos, and arc lamps). More importantly, the intimate relation between electric charge and electromagnetism had been well documented following the discoveries of Michael Faraday and Clerk Maxwell. Since electromagnetism was *known* to be a wave generated by a changing electric or magnetic *field* (a continuous, wave-like entity itself) an atomic/particle description of electricity and charge was a non sequitur. And classical electrodynamics was not the only classical theory rendered incomplete.

Radiation Quantization

Black-body radiation, the emission of electromagnetic energy due to an object's heat, could not be explained from classical arguments alone. The equipartition theorem of classical mechanics, the basis of all classical thermodynamic theories, stated that an object's energy is partitioned equally among the object's vibrational modes. This worked well when describing thermal objects, whose vibrational modes were defined as the speeds of their constituent atoms, and the speed distribution derived from egalitarian partitioning of these vibrational modes closely matched experimental results.

Speeds much higher than the average speed were suppressed by the fact that kinetic energy is quadratic – doubling the speed requires four times the energy – thus the number of atoms occupying high energy modes (high speeds) quickly drops off because the constant, equal partition can excite successively fewer atoms. Low speed modes would *ostensibly* dominate the distribution, since low speed modes would require ever less energy, and *prima facie* a zero-speed mode would require zero energy and its energy partition would contain an infinite number of atoms. *But* this would only occur in the absence of atomic interaction; when collisions are allowed, the low speed modes are immediately suppressed by jostling from the higher energy atoms, exciting them to higher energy modes. An equilibrium is swiftly reached where most atoms occupy a speed proportional to the temperature of the object (thus defining temperature as the average kinetic energy of the object).

But applying the same reasoning to the electromagnetic emission of such a thermal object was not so successful. It had been long known that thermal objects emit light. Hot metal glows red, and upon further heating, white (this is the underlying principle of the incandescent bulb). Since light was known to be waves of electromagnetism, physicists hoped to describe this emission via classical laws. This became known as the black body problem. Since the equipartition theorem worked so well in describing the vibrational modes of the thermal object itself, it was trivial to assume that it would perform equally well in describing the radiative emission of such objects. But a problem quickly arose when determining the vibrational modes of light. To simplify the problem (by limiting the vibrational modes) a lowest allowable wavelength was defined by placing the thermal object in a cavity. Any electromagnetic mode at equilibrium (i.e. any standing wave) could only exist if it used the walls of the cavities as nodes. Thus there were no waves/modes with a wavelength larger than twice the length (L) of the cavity.

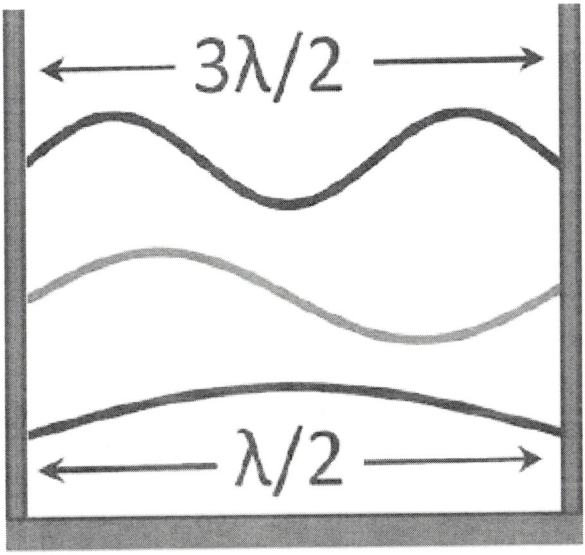

***Figure:** Standing waves in a cavity*

The first few allowable modes would therefore have wavelengths of : $2L$, L, $2L/3$, $L/2$, etc. (each successive wavelength adding one node to the wave). However, while the wavelength could never exceed $2L$, there was no such limit on decreasing the wavelength, and adding nodes to reduce the wavelength could proceed *ad infinitum*. Suddenly it became apparent that the short wavelength modes completely dominated the distribution, since ever shorter wavelength modes could be crammed into the cavity. If each mode received an equal partition of energy, the short wavelength modes would consume all the energy. This became clear when plotting the Rayleigh–Jeans law which, while correctly predicting the intensity of long wavelength emissions, predicted infinite total energy as the intensity diverges to infinity for short wavelengths. This became known as the ultraviolet catastrophe.

The solution arrived in 1900 when Max Planck hypothesized that the frequency of light emitted by the black body depended on the frequency of the *oscillator* that emitted it, and the energy of these oscillators increased linearly with frequency (according to his constant h, where $E = h\nu$).

This was not an unsound proposal considering that macroscopic oscillators operate similarly: when studying five simple harmonic oscillators of equal amplitude but different frequency, the oscillator with the highest frequency possesses the highest energy (though this relationship is not linear like Planck's).

By demanding that high-frequency light must be emitted by an oscillator of equal frequency, and further requiring that this oscillator occupy higher energy than one of a lesser frequency, Planck avoided any catastrophe; giving an equal partition to high-frequency oscillators produced successively fewer oscillators and less emitted light. And as in the Maxwell–Boltzmann distribution, the low-frequency, low-energy oscillators were suppressed by the onslaught of thermal jiggling from higher energy oscillators, which necessarily increased their energy and frequency.

The most revolutionary aspect of Planck's treatment of the black body is that it inherently relies on an integer number of oscillators in thermal equilibrium with the electromagnetic field. These oscillators give their entire energy to the electromagnetic field, creating a quantum of light, as often as they are excited by the electromagnetic field, absorbing a quantum of light and beginning to oscillate at the corresponding frequency. Planck had intentionally created an atomic theory of the black body, but had unintentionally generated an atomic theory of light, where the black body never generates quanta of light at a given frequency with an energy less than $h\nu$. However, once realising that he had quantized the electromagnetic field, he denounced particles of light as a limitation of his approximation, not a property of reality.

The Photoelectric Effect Illuminated

Yet while Planck had solved the ultraviolet catastrophe by using atoms and a quantized electromagnetic field, most physicists immediately agreed that Planck's "light quanta" were unavoidable flaws in his model. A more complete derivation of black body radiation would produce a fully continuous, fully wave-like electromagnetic field with no quantization.

However, in 1905 Albert Einstein took Planck's black body model in itself and saw a wonderful solution to another outstanding problem of the day: the photoelectric effect. Ever since the discovery of electrons eight years previously, electrons had been *the* thing to study in physics laboratories worldwide. Nikola Tesla discovered in 1901 that when a metal was illuminated by high-frequency light (e.g. ultraviolet light), electrons were ejected from the metal at high energy. This work was based on the previous knowledge that light incident upon metals produces a current, but Tesla was the first to describe it as a particle phenomenon.

The following year, Philipp Lenard discovered that (within the range of the experimental parameters he was using) the energy of these ejected electrons did *not* depend on the intensity of the incoming light, but on its *frequency*. So if one shines a little low-frequency light upon a metal, a few low energy electrons are ejected. If one now shines a very intense beam of low-frequency light upon the same metal, a whole slew of electrons are ejected; however they possess the same low energy, there are merely *more of them*. In order to get high energy electrons, one must illuminate the metal with high-frequency light. The more light there is, the more electrons are ejected. Like blackbody radiation, this was at odds with a theory invoking continuous transfer of energy between radiation and matter. However, it can still be explained using a fully classical description of light, as long as matter is quantum mechanical in nature.

If one used Planck's energy quanta, and demanded that electromagnetic radiation at a given frequency could only transfer energy to matter in integer multiples of an energy quantum hv, then the photoelectric effect could be explained very simply. Low-frequency light only ejects low-energy electrons because each electron is excited by the absorption of a single photon. Increasing the intensity of the low-frequency light (increasing the number of photons) only increases the number of excited electrons, not their energy, because the energy of each photon remains low. Only by increasing the frequency of the light, and thus increasing the energy of the photons, can one eject electrons with higher energy. Thus, using Planck's constant h to determine the energy of the photons based upon their frequency, the energy of ejected electrons should also increase linearly with frequency; the gradient of the line being Planck's constant. These results were not confirmed until 1915, when Robert Andrews Millikan, who had previously determined the charge of the electron, produced experimental results in perfect accord with Einstein's predictions.

While the energy of ejected electrons reflected Planck's constant, the existence of photons was not explicitly proven until the discovery of the photon antibunching effect, of which a modern experiment can be performed in undergraduate-level labs. This phenomenon could only be explained via photons, and not through any semi-classical theory (which could alternatively explain the photoelectric effect). When Einstein received his Nobel Prize in 1921, it was not for his more difficult and mathematically labourious special and general relativity, but for the simple, yet totally revolutionary, suggestion of quantized light.

Einstein's "light quanta" would not be called photons until 1925, but even in 1905 they represented the quintessential example of wave–particle duality. Electromagnetic radiation propagates following linear wave equations, but can only be emitted or absorbed as discrete elements, thus acting as a wave and a particle simultaneously.

Developmental Milestones

Huygens and Newton: The earliest comprehensive theory of light was advanced by Christiaan Huygens, who proposed a wave theory of light, and in particular demonstrated how waves might interfere to form a wavefront, propagating in a straight line. However, the theory had difficulties in other matters, and was soon overshadowed by Isaac Newton's corpuscular theory of light. That is, Newton proposed that light consisted of small particles, with which he could easily explain the phenomenon of reflection. With considerably more difficulty, he could also explain refraction through a lens, and the splitting of sunlight into a rainbow by a prism. Newton's particle viewpoint went essentially unchallenged for over a century.

Young, Fresnel, and Maxwell

In the early 19th century, the double-slit experiments by Young and Fresnel provided evidence for Huygens' wave theories. The double-slit experiments showed that when light is sent through a grid, a characteristic interference pattern is observed, very similar to the pattern resulting from the interference of water waves; the wavelength of light can be computed from such patterns. The wave view did not immediately displace the ray and particle view, but began to dominate scientific thinking about light in the mid 19th century, since it could explain polarization phenomena that the alternatives could not.

In the late 19th century, James Clerk Maxwell explained light as the propagation of electromagnetic waves according to the Maxwell equations. These equations were verified by experiment by Heinrich Hertz in 1887, and the wave theory became widely accepted.

Planck's Formula for Black-Body Radiation

In 1901, Max Planck published an analysis that succeeded in reproducing the observed spectrum of light emitted by a glowing object. To accomplish this, Planck had to make an ad hoc mathematical assumption of quantized energy of the oscillators (atoms of the black body) that emit radiation. It was Einstein who later proposed that it is the electromagnetic radiation itself that is quantized, and not the energy of radiating atoms.

Einstein's Explanation of the Photoelectric Effect

In 1905, Albert Einstein provided an explanation of the photoelectric effect, a hitherto troubling experiment that the wave theory of light seemed incapable of explaining. He did so by postulating the existence of photons, quanta of light energy with particulate qualities. In the photoelectric effect, it was observed that shining a light on certain metals would lead to an electric current in a circuit. Presumably, the light was knocking electrons out of the metal, causing current to flow. However, using the case of potassium as an example, it was also observed that while a dim blue light was enough to cause a current, even the strongest, brightest red light available with the technology of the time caused no current at all. According to the classical theory of light and matter, the strength or amplitude of a light wave was in proportion to its brightness: a bright light should have been easily strong enough to create a large current. Yet, oddly, this was not so.

Einstein explained this conundrum by postulating that the electrons can receive energy from electromagnetic field only in discrete portions (quanta that were called photons): an amount of energy E that was related to the frequency f of the light by

$$E = hf$$

where h is Planck's constant (6.626×10^{-34} J seconds). Only photons of a high enough frequency (above a certain *threshold* value) could knock an electron free. For example, photons of blue light had sufficient energy to free an electron from the metal, but photons of red light did not. More intense light above the threshold frequency could release more electrons, but no amount of light (using technology available at the time) below the threshold frequency could release an electron. To "violate" this law would require extremely high intensity lasers which had not yet been invented. Intensity-dependent phenomena have now been studied in detail with such lasers.

De Broglie's Wavelength

In 1924, Louis-Victor de Broglie formulated the de Broglie hypothesis, claiming that *all* matter, not just light, has a wave-like nature; he related wavelength (denoted as λ), and momentum (denoted as p):

$$\lambda = \frac{h}{p}$$

This is a generalisation of Einstein's equation above, since the momentum of a photon is given by $p = \frac{E}{c}$ and the wavelength (in a vacuum) by $\lambda = \frac{c}{f}$, where c is the speed of light in vacuum.

De Broglie's formula was confirmed three years later for electrons (which differ from photons in having arest mass) with the observation of electron diffraction in two independent experiments. At the University of Aberdeen, George Paget Thomson passed a beam of electrons through a thin metal film and observed the predicted interference patterns. At Bell Labs Clinton Joseph Davisson and Lester Halbert Germer guided their beam through a crystalline grid.

De Broglie was awarded the Nobel Prize for Physics in 1929 for his hypothesis. Thomson and Davisson shared the Nobel Prize for Physics in 1937 for their experimental work.

Heisenberg's Uncertainty Principle

In his work on formulating quantum mechanics, Werner Heisenberg postulated his uncertainty principle, which states:

$$\Delta x \Delta p \geq \frac{\hbar}{2}$$

where

> Δ here indicates standard deviation, a measure of spread or uncertainty;
>
> x and p are a particle's position and linear momentum respectively.
>
> $\hbar$ is the reduced Planck's constant (Planck's constant divided by 2π).

Heisenberg originally explained this as a consequence of the process of measuring: Measuring position accurately would disturb momentum and vice-versa, offering an example (the "gamma-ray microscope") that depended crucially on the de Broglie hypothesis. It is now thought, however, that this only partly explains the phenomenon, but that the uncertainty also exists in the particle itself, even before the measurement is made.

In fact, the modern explanation of the uncertainty principle, extending the Copenhagen interpretation first put forward by Bohr and Heisenberg, depends even more centrally on the wave nature of a particle: Just as it is nonsensical to discuss the precise location of a wave on a string, particles do not have perfectly precise positions; likewise, just as it is nonsensical to discuss the wavelength of a "pulse" wave travelling down a string, particles do not have perfectly precise momenta (which corresponds to the inverse of wavelength).

Moreover, when position is relatively well defined, the wave is pulse-like and has a very ill-defined wavelength (and thus momentum). And conversely, when momentum (and thus wavelength) is relatively well defined, the wave looks long and sinusoidal, and therefore it has a very ill-defined position.

De Broglie himself had proposed a pilot wave construct to explain the observed wave–particle duality. In this view, each particle has a well-defined position and momentum, but is guided by a wave function derived from Schrödinger's equation.

The pilot wave theory was initially rejected because it generated non-local effects when applied to systems involving more than one particle. Non-locality, however, soon became established as an integral feature of quantum theory (EPR paradox), and David Bohm extended de Broglie's model to explicitly include it. In the resulting representation, also called the de Broglie–Bohm theory or Bohmian mechanics, the wave–particle duality is not a property of matter itself, but an appearance generated by the particle's motion subject to a guiding equation or quantum potential.

Wave Behaviour of Large Objects

Since the demonstrations of wave-like properties in photons and electrons, similar experiments have been conducted with neutrons and protons. Among the most famous experiments are those of Estermann and Otto Stern in 1929. Authors of similar recent experiments with atoms and molecules, described below, claim that these larger particles also act like waves.

A dramatic series of experiments emphasizing the action of gravity in relation to wave–particle duality were conducted in the 1970s using the neutron interferometre. Neutrons, one of the components of the atomic nucleus, provide much of the mass of a nucleus and thus of ordinary matter.

In the neutron interferometre, they act as quantum-mechanical waves directly subject to the force of gravity. While the results were not surprising since gravity was known to act on everything, including light (tests of general relativity and the Pound-Rebka falling photon experiment), the self-interference of the quantum mechanical wave of a massive fermion in a gravitational field had never been experimentally confirmed before.

In 1999, the diffraction of C_{60} fullerenes by researchers from the University of Vienna was reported. Fullerenes are comparatively large and massive objects, having an atomic mass of about 720 u. Thede Broglie wavelength is 2.5 pm, whereas the diametre of the molecule is about 1 nm, about 400 times larger. As of 2005, this is the largest object for which quantum-mechanical wave-like properties have been directly observed in far-field diffraction.

In 2003 the Vienna group also demonstrated the wave nature of tetraphenylporphyrin—a flat biodye with an extension of about 2 nm and a mass of 614 u.

For this demonstration they employed a near-field Talbot Lau interferometre. In the same interferometre they also found interference fringes for $C_{60}F_{48.}$, a fluorinated buckyball with a mass of about 1600 u, composed of 108 atoms. Large molecules are already so complex that they give experimental access to some aspects of the quantum-classical interface, i.e. to certain decoherence mechanisms.

Whether objects heavier than the Planck mass (about the weight of a large bacterium) have a de Broglie wavelength is theoretically unclear and experimentally unreachable; above the Planck mass a particle's Compton wavelength would be smaller than the Planck length and its own Schwarzschild radius, a scale at which current theories of physics may break down or need to be replaced by more general ones.

Recently Couder, Fort *et al.* showed that we can use macroscopic oil droplets on vibrating surface as a model of wave-particle duality - localised droplet creates periodical waves around and interaction with them leads to quantum-like phenomena: interference in double-slit experiment, unpredictable tunnelling (depending in complicated way on practically hidden state of field) and orbit quantization (that particle has to 'find a resonance' with field perturbations it creates - after one orbit, its internal phase has to return to the initial state).

Treatment in Modern Quantum Mechanics

Wave–particle duality is deeply embedded into the foundations of quantum mechanics, so well that modern practitioners rarely discuss it as such.

In the formalism of the theory, all the information about a particle is encoded in its *wave function*, a complex valued function roughly analogous to the amplitude of a wave at each point in space. This function evolves according to a differential equation (generically called the Schrödinger equation), and this equation gives rise to wave-like phenomena such as interference and diffraction.

The particle-like behaviour is most evident due to phenomena associated with measurement in quantum mechanics.

Upon measuring the location of the particle, the wave-function will randomly "collapse," or rather, "decoheres" to a sharply peaked function at some location, with the likelihood of any particular location equal to the squared amplitude of the wave-function there.

The measurement will return a well-defined position, (subject to uncertainty), a property traditionally associated with particles.

Although this picture is somewhat simplified (to the non-relativistic case), it is adequate to capture the essence of current thinking on the phenomena historically called "wave–particle duality".

Illustration of wave-particle duality in one dimension for one particle - probability of finding a particle in space is distributed as a (complex valued) waveform through space, mathematically described by the particle's wavefunction ψ. The colour opacity (%) of the particles corresponds to the probability density of finding the particle at the points on the x-axis (waves correspond to ψ, particles location to $\psi^* \psi$).

The wave-particle model is consistent with *De Broglie's hypothesis* and *Heisenberg's uncertainty principle*. Note that uncertainty in wavelength $\Delta\lambda$ is not the same as uncertainty in position Δx.

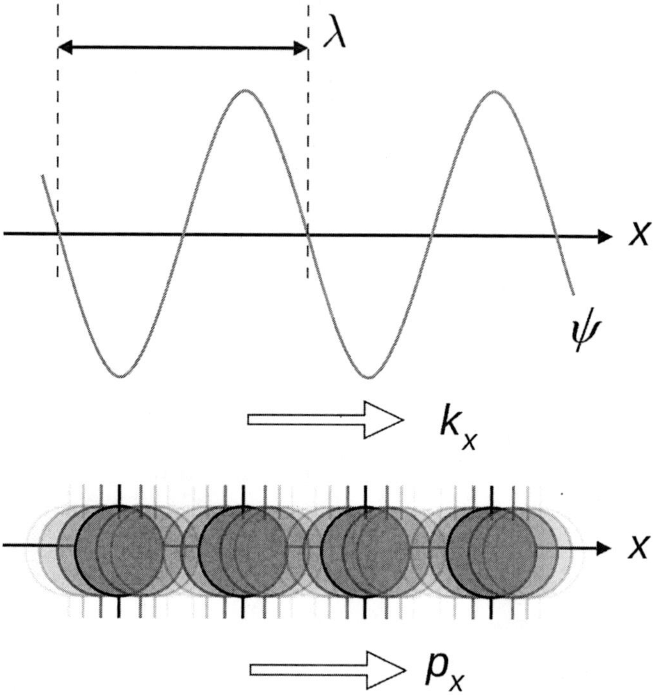

***Figure:** Wave-particle with a measurable wavelength λ has momentum p, wave-vector k and energy E. The uncertainty in momentum Δp or energy ΔE is small. The wave-like probability distribution ψ does not specify a single place where the particle is, due to the periodicity of the wave, any of the amplitude (or antinode) locations are the most likely regions the particle is at time t and position x, so the uncertainty in position Δx or time Δt are both large. This fits in with the Uncertainty principle.*

Alternative Views

An editor has expressed a concern that this article lends undue weight to certain ideas, incidents, controversies or matters relative to the article subject as a whole. Please help to create a more balanced presentation. Discuss and resolve this issue before removing this message. *(July 2009)*

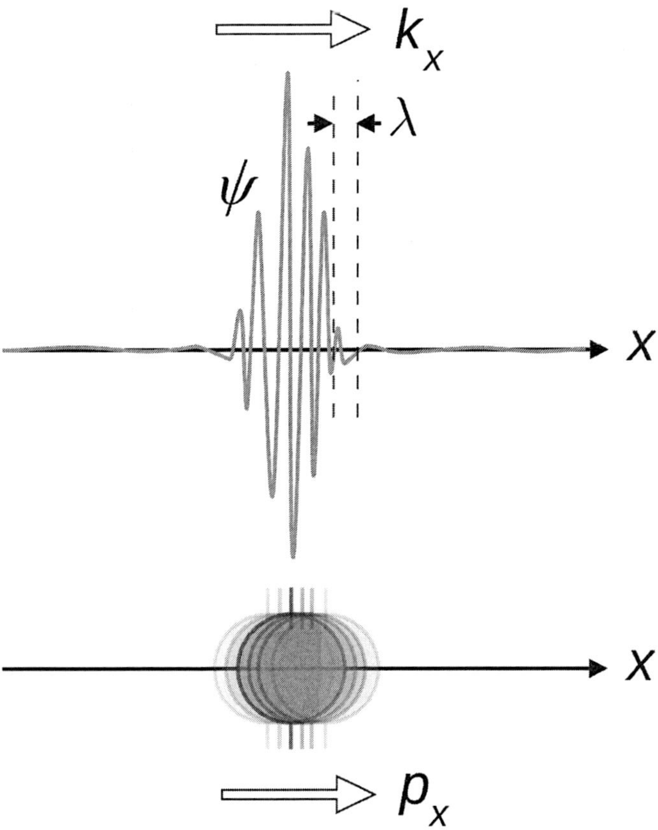

Figure: *Inverse of the previous situation - if λ is unknown, so is* p, k, *and* E, *but* ψ *is clearer in* x *and* t. *So this time* Δx *or* Δt *is smaller than* Δp *or* ΔE.

Particle-Only View

The pilot wave model, originally developed by Louis de Broglie and further developed by David Bohminto the hidden variable theory proposes that there is no duality, but rather particles are guided, in adeterministic fashion, by a pilot wave (or its "quantum potential") which will direct them to areas of constructive interference in preference to areas of destructive interference. This idea is held by a significant minority within the physics community.

At least one physicist considers the "wave-duality" a misnomer, as L. Ballentine, *Quantum Mechanics, A Modern Development*, p. 4, explains:

When first discovered, particle diffraction was a source of great puzzlement. Are "particles" really "waves?" In the early experiments, the diffraction patterns were detected holistically by means of a photographic plate, which could not detect individual particles. As a result, the notion grew that particle and wave properties were mutually incompatible, or complementary, in the sense that different measurement apparatuses would be required to observe them.

That idea, however, was only an unfortunate generalisation from a technological limitation. Today it is possible to detect the arrival of individual electrons, and to see the diffraction pattern emerge as a statistical pattern made up of many small spots (Tonomura et al., 1989).

Evidently, quantum particles are indeed particles, but whose behaviour is very different from classical physics would have us to expect.

Afshar's experiment (2007) has demonstrated that it is possible to simultaneously observe both wave and particle properties of photons. Biddulph (2010) has explained this by applying techniques from deterministic chaos to non-chaotic systems, in particular a computable version of Palmer's Universal Invariant Set proposition (2009), which allows the apparent weirdness of quantum phenomena to be explained as artefacts of the quantum apparatus not a fundamental property of nature.

Waves are shown to be the only means of describing motion, since smooth motion on a continuum is impossible. If a particle visits every point on its trajectory then the motion is an algorithm for each point. Turing has shown that almost all numbers are non-computable, which means that there is no possible algorithm, so the set of points on a trajectory is sparse.

This implies that motion is either jerky or wave-like. By removing the need to load the particle with the properties of space and time, a fully deterministic, local and causal description of quantum phenomena is possible by use of a simple dynamical operator on a Universal Invariant Set.

Wave-Only View

At least one scientist proposes that the duality can be replaced by a "wave-only" view. Carver Mead's *Collective Electrodynamics: Quantum Foundations of Electromagnetism* (2000) analyses the behaviour of electrons and photons purely in terms of electron wave functions, and attributes the apparent particle-like behaviour to quantization effects and eigenstates. According to reviewer David Haddon:

Mead has cut the Gordian knot of quantum complementarity. He claims that atoms, with their neutrons, protons, and electrons, are not particles at all but pure waves of matter.

Mead cites as the gross evidence of the exclusively wave nature of both light and matter the discovery between 1933 and 1996 of ten examples of pure wave phenomena, including the ubiquitous laser of CD players, the self-propagating electrical currents of superconductors, and the Bose–Einstein condensate of atoms.

Albert Einstein, who, in his search for a Unified Field Theory, did not accept wave-particle duality, wrote:

This double nature of radiation (and of material corpuscles)...has been interpreted by quantum-mechanics in an ingenious and amazingly successful fashion. This interpretation...appears to me as only a temporary way out...

And theoretical physicist Mendel Sachs, who completed Einstein's unified field theory, writes:

Instead, one has a single, holistic continuum, wherein what were formerly called discrete, separable particles of matter are instead the infinite number of distinguishable, though correlated manifestations of this continuum, that in principle is the universe. Hence, wave-particle dualism, which is foundational for the quantum theory, is replaced by wave (continuous field) monism.

The many-worlds interpretation (MWI) is sometimes presented as a waves-only theory, including by its originator, Hugh Everett who referred to MWI as "the wave interpretation".

The *Three Wave Hypothesis* of R. Horodecki relates the particle to wave. The hypothesis implies that a massive particle is an intrinsically spatially as well as temporally extended wave phenomenon by a nonlinear law. According to M. I. Sanduk this hypothesis is related to a hypothetical bevel gear model. Then both concepts of particle and wave may be attributed to an observation problem of the gear.

Neither-Wave-nor-Particle View

It has been argued that there are never exact particles or waves, but only some compromise or intermediate between them. One consideration is that zero dimensional mathematical points cannot be observed. Another is that the formal representation of such points, the Kronecker delta function is unphysical, because it cannot be renormalized.

Parallel arguments apply to pure wave states.

> *"Such positions states are idealised wavefunctions [..] Whereas the momentum states are infinitely spread out, the position states are infinitely concentrated. Neither is normaliseable[..]"*

Relational Approach to Wave–Particle Duality

Relational quantum mechanics is developed which regards the detection event as establishing a relationship between the quantized field and the detector.

The inherent ambiguity associated with applying Heisenberg's uncertainty principle and thus wave–particle duality is subsequently avoided.

Applications

Although it is difficult to draw a line separating wave–particle duality from the rest of quantum mechanics, it is nevertheless possible to list some applications of this basic idea.

- Wave–particle duality is exploited in electron microscopy, where the small wavelengths associated with the electron can be used to view objects much smaller than what is visible using visible light.
- Similarly, neutron diffraction uses neutrons with a wavelength of about 0.1 nm, the typical spacing of atoms in a solid, to determine the structure of solids.

Chapter 6

Bohr's Model

In 1913 Niels Bohr proposed a new model of the atom that included quantized electron orbits. In Bohr's model, electrons could inhabit only certain orbits around the atomic nucleus. When an atom emitted (or absorbed) energy, the electron did not move in a continuous trajectory from one orbit around the nucleus to another, as might be expected classically. Instead, the electron would jump instantaneously from one orbit to another, giving off the emitted light in the form of a photon. The possible energies of photons given off by each element were determined by the differences in energy between the orbits, and so the emission spectrum for each element would contain a number of lines.

Bohr theorised that the angular momentum, L, of an electron is quantised:

$$L = n\frac{h}{2\pi},$$

where n is an integer and h is the Planck constant. Starting from this assumption, Coulomb's law and the equations of circular motion show that an electron with n units of angular momentum will orbit a proton at a distance r given by

$$r = \frac{n^2h^2}{4\pi^2k_e me^2}$$

where k_e is the Coulomb constant, m is the mass of an electron, and e is the charge on an electron. For simplicity this is written as

$$r = n^2a_0,$$

where a_0, called the Bohr radius, is equal to 0.0529 nm. The Bohr radius is the radius of the smallest allowed orbit.

The energy of the electron can also be calculated, and is given by

$$E = -\frac{k_e e^2}{2a_0}\frac{1}{n^2}$$

Thus Bohr's assumption that angular momentum is quantised means that an electron can only inhabit certain orbits around the nucleus, and that it can have only certain energies. A consequence of these constraints is that the electron will not crash into the nucleus: it cannot continuously emit energy, and it cannot come closer to the nucleus than a_0 (the Bohr radius).

An electron loses energy by jumping instantaneously from its original orbit to a lower orbit; the extra energy is emitted in the form of a photon. Conversely, an electron that absorbs a photon gains energy, hence it jumps to an orbit that is farther from the nucleus.

Each photon from glowing atomic hydrogen is due to an electron moving from a higher orbit, with radius r_n, to a lower orbit, r_m. The energy E_n of this photon is the difference in the energies E_n and E_m of the electron:

$$E_\gamma = E_n - E_m = \frac{k_e e^2}{2a_0}\left(\frac{1}{m^2} - \frac{1}{n^2}\right)$$

Since Planck's equation shows that the photon's energy is related to its wavelength by $E_\gamma = hc/\lambda$, the wavelengths of light which can be emitted are given by

$$\frac{1}{\lambda} = \frac{k_e e^2}{2a_0 hc}\left(\frac{1}{m^2} - \frac{1}{n^2}\right).$$

This equation has the same form as the Rydberg formula, and predicts that the constant R should be given by

$$R = \frac{k_e e^2}{2a_0 hc}.$$

Therefore the Bohr model of the atom can predict the emission spectrum of hydrogen in terms of fundamental constants. However, it was not able to make accurate predictions for multi-electron atoms, or to explain why some spectral lines are brighter than others.

Wave-Particle Duality

In 1924, Louis de Broglie proposed the idea that just as light has both wave-like and particle-like properties, matter also has wave-like properties. The wavelength, λ , associated with a particle is related to ts momentum, p:

$$p = \frac{h}{\lambda}.$$

The relationship, called the de Broglie hypothesis, holds for all types of matter. Thus all matter exhibits properties of both particles and waves.

Three years later, the wave-like nature of electrons was demonstrated by showing that a beam of electrons could exhibit diffraction, just like a beam of light. At the University of Aberdeen, George Thomson passed a beam of electrons through a thin metal film and observed the predicted interference patterns. At Bell Labs, Davisson and Germer guided their beam through a crystalline grid. Similar wave-like phenomena were later shown for atoms and even small molecules. De Broglie was awarded the Nobel Prize for Physics in 1929 for his hypothesis; Thomson and Davisson shared the Nobel Prize for Physics in 1937 for their experimental work.

This is the concept of wave-particle duality: neither the classical concepts of "particle" or "wave" can fully describe the behaviour of quantum-scale objects, either photons or matter. Wave-particle duality is an example of the principle of complementarity in quantum physics. An elegant example of wave-particle duality, the double slit experiment, is discussed in the section below.

De Broglie's treatment of quantum events served as a jumping off point for Schrödinger when he set about to construct a wave equation to describe quantum theoretical events.

The Double-Slit Experiment

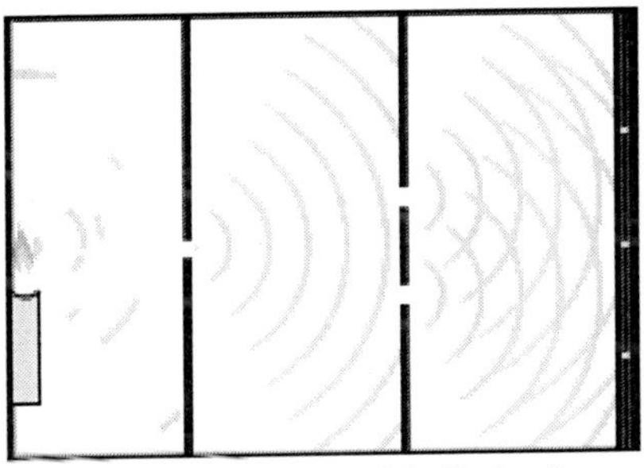

Figure: *Light from one slit interferes with light from the other, producing an interference pattern (the 3 fringes shown at the right).*

In the double-slit experiment as originally performed by Thomas Young and Augustin Fresnel in 1827, a beam of light is directed through two narrow, closely spaced slits, producing an interference pattern of light and dark bands on a screen. If one of the slits is covered up, one might naively expect that the intensity of the fringes due to interference would be halved everywhere. In fact, a much simpler pattern is seen, a simple diffraction pattern. Closing one slit results in a much simpler pattern diametrically opposite the open slit. Exactly the same behaviour can be demonstrated in water waves, and so the double-slit experiment was seen as a demonstration of the wave nature of light.

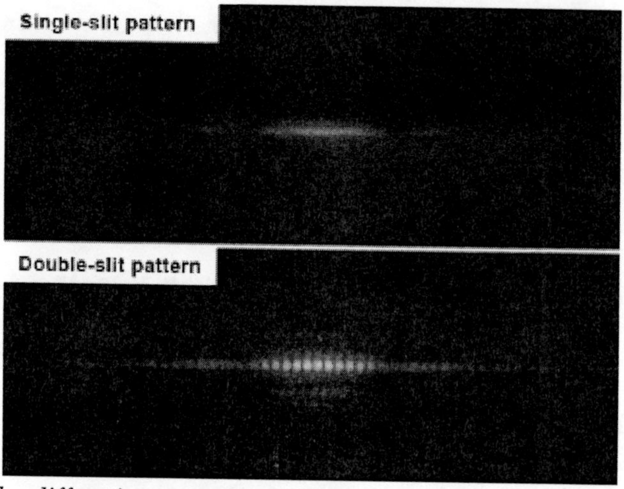

Figure: *The diffraction pattern produced when light is shone through one slit (top) and the interference pattern produced by two slits (bottom). The interference pattern from two slits is much more complex, demonstrating the wave-like propagation of light.*

The double-slit experiment has also been performed using electrons, atoms, and even molecules, and the same type of interference pattern is seen. Thus all matter possesses both particle and wave characteristics.

Even if the source intensity is turned down so that only one particle (e.g. photon or electron) is passing through the apparatus at a time, the same interference pattern develops over time. The quantum particle acts as a wave when passing through the double slits, but as a particle when it is detected. This is a typical feature of quantum complementarity: a quantum particle will act as a wave when we do an experiment to measure its wave-like properties, and like a particle when we do an experiment to measure its particle-like properties. Where on the detector screen any individual particle shows up will be the result of an entirely random process.

Application to the Bohr Model

De Broglie expanded the Bohr model of the atom by showing that an electron in orbit around a nucleus could be thought of as having wave-like properties. In particular, an electron will be observed only in situations that permit a standing wave around a nucleus. An example of a standing wave is a violin string, which is fixed at both ends and can be made to vibrate. The waves created by a stringed instrument appear to oscillate in place, moving from crest to trough in an up-and-down motion. The wavelength of a standing wave is related to the length of the vibrating object and the boundary conditions. For example, because the violin string is fixed at both ends, it can carry standing waves of wavelengths 2l/n, where l is the length and n is a positive integer. De Broglie suggested that the allowed electron orbits were those for which the circumference of the orbit would be an integer number of wavelengths.

Development of Modern Quantum Mechanics

In 1925, building on de Broglie's hypothesis, Erwin Schrödinger developed the equation that describes the behaviour of a quantum mechanical wave. The equation, called the Schrödinger equation after its creator, is central to quantum mechanics, and defines the permitted stationary states of a quantum system, and describes how the quantum state of a physical system changes in time.

Schrödinger was able to calculate the energy levels of hydrogen by treating a hydrogen atom's electron as a classical wave, moving in a well of electrical potential created by the proton. This calculation accurately reproduced the energy levels of the Bohr model.

At around the same time, Werner Heisenberg was trying to find an explanation for the intensities of the different lines in the hydrogen emission spectrum. By means of a series of mathematical analogies, Heisenberg wrote out the quantum mechanical analogue for the classical computation of intensities. Shortly afterwards, Heisenberg's colleague Max Born realised that Heisenberg's method of calculating the probabilities for transitions between the different energy levels could best be expressed by using the mathematical concept of matrices.

In May 1926, Schrödinger proved that Heisenberg's matrix mechanics and his own wave mechanics made the same predictions about the properties and behaviour of the electron; mathematically, the two theories were identical. Yet the two men disagreed on the interpretation of their mutual theory. Heisenberg saw no problem in the existence of discontinuous quantum jumps, but Schrödinger hoped that a theory based on continuous wave-like properties could avoid what he called (in the words of Wilhelm Wien) "this nonsense about quantum jumps."

Copenhagen Interpretation

Bohr, Heisenberg and others tried to explain what these experimental results and mathematical models really mean. Their description, known as the Copenhagen interpretation of quantum mechanics, aimed to describe the nature of reality that was being probed by the measurements and described by the mathematical formulations of quantum mechanics.

The main principles of the Copenhagen interpretation are:

1. A system is completely described by a wave function, ψ. (Heisenberg)
2. How ψ changes over time is given by the Schrödinger equation.
3. The description of nature is essentially probabilistic. The probability of an event — for example, where on the screen a particle will show up in the two slit experiment — is related to the square of the amplitude of its wave function. (Born rule, due to Max Born, which gives a physical meaning to the wavefunction in the Copenhagen interpretation: the probability amplitude)
4. It is not possible to know the values of all of the properties of the system at the same time; those properties that are not known with precision must be described by probabilities. (Heisenberg's uncertainty principle)
5. Matter, like energy, exhibits a wave-particle duality. An experiment can demonstrate the particle-like properties of matter, or its wave-like properties; but not both at the same time. (Complementarity principle due to Bohr)
6. Measuring devices are essentially classical devices, and measure classical properties such as position and momentum.
7. The quantum mechanical description of large systems should closely approximate the classical description. (Correspondence principle of Bohr and Heisenberg)

Various consequences of these principles are discussed in more detail in the following subsections.

Uncertainty Principle

Suppose that we want to measure the position and speed of an object — for example a car going through a radar speed trap. Naively, we assume that the car has a definite position and speed at a particular moment in time, and how accurately we can measure these values depends on the quality of our measuring equipment — if we improve the precision of our measuring equipment, we will get a result that is closer to the true value. In particular, we would assume that how precisely we measure the speed of the car does not affect its position, and vice versa.

In 1927, Heisenberg proved that these assumptions are not correct. Quantum mechanics shows that certain pairs of physical properties, like position and speed, cannot both be known to arbitrary precision: the more precisely one property is known, the less precisely the other can be known. This statement is known as the uncertainty principle. The uncertainty principle isn't a statement about the accuracy of our measuring equipment, but about the nature of the system itself — our naive assumption that the car had a definite position and speed was incorrect. On a scale of cars and people, these uncertainties are too small to notice, but when dealing with atoms and electrons they become critical.

The uncertainty principle shows mathematically that the product of the uncertainty in the position and momentum of a particle (momentum is velocity multiplied by mass) could never be less than a certain value, and that this value is related to Planck's constant.

Wave Function Collapse

Wavefunction collapse is a forced term for whatever happened when it becomes appropriate to replace the description of an uncertain state of a system by a description of the system in a definite state. Explanations for the nature of the process of becoming certain are controversial. At any time before a photon "shows up" on a detection screen it can only be described by a set of probabilities for where it might show up. When it does show up, for instance in the CCD of an electronic camera, the time and the space where it interacted with the device are known within very tight limits. However, the photon has disappeared, and the wave function has disappeared with it. In its place some physical change in the detection screen has appeared, e.g., an exposed spot in a sheet of photographic film.

Eigenstates and Eigenvalues

Because of the uncertainty principle, statements about both the position and momentum of particles can only assign a probability that the position or momentum will have some numerical value. Therefore it is necessary to formulate clearly the difference between the state of something that is indeterminate, such as an electron in a probability cloud, and the state of something having a definite value. When an object can definitely be "pinned-down" in some respect, it is said to possess an eigenstate.

The Pauli Exclusion Principle

Wolfgang Pauli proposed the following concise statement of his 1925 principle: "There cannot exist an atom in such a quantum state that two electrons within [it] have the same set of quantum numbers."

He developed the exclusion principle from what he called a "two-valued quantum degree of freedom" to account for the observation of a doublet, meaning a pair of lines differing by a small amount (e.g. on the order of 0.15 Å), in the spectrum of atomic hydrogen. The existence of these closely spaced lines in the bright-line spectrum meant that there was more energy in the electron orbital from magnetic moments than had previously been described.

In early 1925, Uhlenbeck and Goudsmit proposed that electrons rotate about an axis in the same way that the earth rotates on its axis. They proposed to call this property spin. Spin would account for the missing magnetic moment, and allow two electrons in the same orbital to occupy distinct quantum states if they, "spun" in opposite directions, thus satisfying the Exclusion Principle. A new quantum number was then needed, one to represent the momentum embodied in the rotation of each electron.

Application to the Hydrogen Atom

Schrödinger was able to calculate the energy levels of hydrogen by treating a hydrogen atom's electron as a wave, represented by the "wave function" ψ , in a electric potential well, V, created by the proton. In Bohr's theory, electrons orbited the nucleus much as planets orbit the sun. Schrödinger's model gives the probability of the electron being located at any particular location – following the uncertainty principle, the electron cannot be viewed as having an exact location at any given time. The solutions to Schrödinger's equation are called atomic orbitals, "clouds" of possible locations in which an electron might be found, a distribution of probabilities rather than a precise location. Orbitals have a range of different shapes in three dimensions. The energies of the different orbitals can be calculated, and they accurately reproduce the energy levels of the Bohr model.

Within Schrödinger's picture, each electron has four properties:

1. An "orbital" designation, indicating whether the particle wave is one that is closer to the nucleus with less energy or one that is farther from the nucleus with more energy;
2. The "shape" of the orbital, spherical or otherwise;
3. The "inclination" of the orbital, determining the magnetic moment of the orbital around the **z**-axis.
4. The "spin" of the electron.

The collective name for these properties is the quantum state of the electron. The quantum state can be described by giving a number to each of these properties; these are known as the electron's quantum numbers. The quantum state of the electron is described by its wavefunction. The Pauli exclusion principle demands that no two electrons within an atom may have the same values of all four numbers.

The first property describing the orbital is the principal quantum number, n, which is the same as in Bohr's model. n denotes the energy level of each orbital. The possible values for n are integers: $n = 1,2,3\ldots$

The next quantum number, the azimuthal quantum number, denoted l, describes the shape of the orbital. The shape is a consequence of the angular momentum of the orbital. The angular momentum represents the resistance of a spinning object to speeding up or slowing down under the influence of external force. The azimuthal quantum number represents the orbital angular momentum of an electron around its nucleus. The possible values for l are integers from 0 to n - 1: $l = 0,1,\ldots,n-1$. The shape of each orbital has its own letter as well. The first shape is denoted by the letter s (a mnemonic being "sphere"). The next shape is denoted by the letter p and has the form of a dumbbell. The other orbitals have more complicated shapes, and are denoted by the letters d, f, and g.

The third quantum number, the magnetic quantum number, describes the magnetic moment of the electron, and is denoted by m_l (or simply m). The possible values for m_l are integers from -l to l: $m_l = -l,-(l-1),\ldots,0,1,\ldots,l$. The magnetic quantum number measures the component of the angular momentum in a particular direction. The choice of direction is arbitrary, conventionally the z-direction is chosen.

The fourth quantum number, the spin quantum number (pertaining to the "orientation" of the electron's spin) is denoted m_s, with values $+ D_2$ or $- D_2$.

The Chemist Linus Pauling Wrote, by Way of Example: In the case of a helium atom with two electrons in the 1 s orbital, the Pauli Exclusion Principle requires that the two electrons differ in the value of one quantum number. Their values of n, l, and m_l are the same; moreover, they have the same spin, $s = D_2$. Accordingly they must differ in the value of m_s, which can have the value of $+ D_2$ for one electron and " D_2 for the other."

It is the underlying structure and symmetry of atomic orbitals, and the way that electrons fill them, that determines the organisation of the periodic table and the structure and strength of chemical bonds between atoms.

Dirac Wave Equation

In 1928, Paul Dirac extended the Pauli equation, which described spinning electrons, to account for special relativity. The result was a theory that dealt properly with events, such as the speed at which an electron orbits the nucleus, occurring at a substantial fraction of the speed of light. By using the simplest electromagnetic interaction, Dirac was able to predict the value of the magnetic moment associated with the electron's spin, and found the experimentally observed value, which was too large to be that of a spinning charged sphere governed by classical physics. He was able to solve for the spectral lines of the hydrogen atom, and to reproduce from physical first principles Sommerfeld's successful formula for the fine structure of the hydrogen spectrum.

Dirac's equations sometimes yielded a negative value for energy, for which he proposed a novel solution: he posited the existence of an antielectron and of a dynamical vacuum. This led to many-particle quantum field theory.

Quantum Entanglement

The Pauli exclusion principle says that two electrons in one system cannot be in the same state. Nature leaves open the possibility, however, that two electrons can have both states "superimposed" over them. Recall that the wave functions that emerge simultaneously from the double slits arrive at the detection screen in a state of superposition.

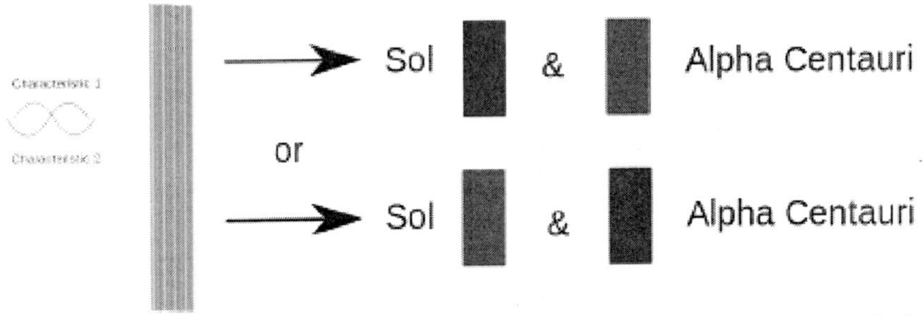

Figure: *Superposition of two quantum characteristics, and two resolution possibilities.*

Nothing is certain until the superimposed waveforms "collapse," At that instant an electron shows up somewhere in accordance with the probabilities that are the squares of the amplitudes of the two superimposed waveforms. The situation there is already very abstract. A concrete way of thinking about entangled photons, photons in which two contrary states are superimposed on each of them in the same event, is as follows:

Imagine that the superposition of a state that can be mentally labelled as blue and another state that can be mentally labelled as red will then appear (in imagination, of course) as a purple state. Two photons are produced as the result of the same atomic event. Perhaps they are produced by the excitation of a crystal that characteristically absorbs a photon of a certain frequency and emits two photons of half the original frequency.

So the two photons come out "purple." If the experimenter now performs some experiment that will determine whether one of the photons is either blue or red, then that experiment changes the photon involved from one having a superposition of "blue" and "red" characteristics to a photon that has only one of those characteristics. The problem that Einstein had with such an imagined situation was that if one of these photons had been kept bouncing between mirrors in a laboratory on earth, and the other one had travelled halfway to the nearest star, when its twin was made to reveal itself as either blue or red, that meant that the distant photon now had to lose its "purple" status too. So whenever it might be investigated, it would necessarily show up, instantaneously, in the opposite state to whatever its twin had revealed.

In trying to show that quantum mechanics was not a complete theory, Einstein started with the theory's prediction that two or more particles that have interacted in the past can appear strongly correlated when their various properties are later measured. He sought to explain this seeming interaction in a classical way, through their common past, and preferably not by some "spooky action at a distance." The argument is worked out in a famous paper, Einstein, Podolsky, and Rosen (1935; abbreviated EPR), setting out what is now called the EPR paradox. Assuming what is now usually called local realism, EPR attempted to show from quantum theory that a particle has both position and momentum simultaneously, while according to the Copenhagen interpretation, only one of those two properties actually exists and only at the moment that it is being measured. EPR concluded that quantum theory is incomplete in that it refuses to consider physical properties which objectively exist in nature. (Einstein, Podolsky, & Rosen 1935 is currently Einstein's most cited publication in physics journals.) In the same year, Erwin Schrödinger used the word "entanglement" and declared: "I would not call that one but rather the characteristic trait of quantum mechanics." The question of whether entanglement is a real condition is still in dispute. The Bell inequalities are the most powerful challenge to Einstein's claims.

Quantum Chemistry

Quantum chemistry is a branch of chemistry whose primary focus is the application of quantum mechanics in physical models and experiments of chemical systems. It involves heavy interplay of experimental and theoretical methods:

- Experimental quantum chemists rely heavily on spectroscopy, through which information regarding the quantization of energy on a molecular scale can be obtained. Common methods are infra-red (IR) spectroscopy and nuclear magnetic resonance (NMR) spectroscopy.
- Theoretical quantum chemistry, the workings of which also tend to fall under the category of computational chemistry, seeks to calculate the predictions of quantum theory; as this task, when applied to polyatomic species, invokes the many-body problem, these calculations are performed using computers rather than by back-of-the-envelope calculations.

In these ways, quantum chemists investigate chemical phenomena.

- In reactions, quantum chemistry studies the ground state of individual atoms and molecules, the excited states, and the transition states that occur during chemical reactions.
- On the calculations: quantum chemical studies use also semi-empirical and other methods based on quantum mechanical

principles, and deal with time dependent problems. Many quantum chemical studies assume the nuclei are at rest (Born-Oppenheimer approximation). Many calculations involve iterative methods that include self-consistent field methods. Major goals of quantum chemistry include increasing the accuracy of the results for small molecular systems, and increasing the size of large molecules that can be processed, which is limited by scaling considerations—the computation time increases as a power of the number of atoms.

History

The history of quantum chemistry essentially began with the 1838 discovery of cathode rays by Michael Faraday, the 1859 statement of the black body radiation problem by Gustav Kirchhoff, the 1877 suggestion by Ludwig Boltzmann that the energy states of a physical system could be discrete, and the 1900 quantum hypothesis by Max Planck that any energy radiating atomic system can theoretically be divided into a number of discrete energy elements └ such that each of these energy elements is proportional to the frequency └ with which they each individually radiate energy, as defined by the following formula:

$$\epsilon = hv$$

where h is a numerical value called Planck's Constant. Then, in 1905, to explain the photoelectric effect (1839), i.e., that shining light on certain materials can function to eject electrons from the material, Albert Einstein postulated, based on Planck's quantum hypothesis, that light itself consists of individual quantum particles, which later came to be called photons (1926). In the years to follow, this theoretical basis slowly began to be applied to chemical structure, reactivity, and bonding.

Electronic Structure

The first step in solving a quantum chemical problem is usually solving the Schrödinger equation (or Dirac equation in relativistic quantum chemistry) with the electronic molecular Hamiltonian. This is called determining the electronic structure of the molecule. It can be said that the electronic structure of a molecule or crystal implies essentially its chemical properties. An exact solution for the Schrödinger equation can only be obtained for the hydrogen atom. Since all other atomic, or molecular systems, involve the motions of three or more "particles", their Schrödinger equations cannot be solved exactly and so approximate solutions must be sought.

Wave Model

The foundation of quantum mechanics and quantum chemistry is the wave model, in which the atom is a small, dense, positively charged nucleus surrounded by electrons. Unlike the earlier Bohr model of the atom, however, the wave model describes electrons as "clouds" moving in orbitals, and their positions are represented by probability distributions rather than discrete points. The strength of this model lies in its predictive power. Specifically, it predicts the pattern of chemically similar elements found in the periodic table. The wave model is so named because electrons exhibit properties (such as interference) traditionally associated with waves.

Valence Bond

Although the mathematical basis of quantum chemistry had been laid by Schrödinger in 1926, it is generally accepted that the first true calculation in quantum chemistry was that of the German physicists Walter Heitler and Fritz London on the hydrogen (H_2) molecule in 1927. Heitler and London's method was extended by the American theoretical physicist John C. Slater and the American theoretical chemist Linus Pauling to become the Valence-Bond (VB) [or Heitler-London-Slater-Pauling (HLSP)] method. In this method, attention is primarily devoted to the pairwise interactions between atoms, and this method therefore correlates closely with classical chemists' drawings of bonds.

Molecular Orbital

An alternative approach was developed in 1929 by Friedrich Hund and Robert S. Mulliken, in which electrons are described by mathematical functions delocalised over an entire molecule. The Hund-Mulliken approach or molecular orbital (MO) method is less intuitive to chemists, but has turned out capable of predicting spectroscopic properties better than the VB method. This approach is the conceptional basis of the Hartree-Fock method and further post Hartree-Fock methods.

Density Functional Theory

The Thomas-Fermi model was developed independently by Thomas and Fermi in 1927. This was the first attempt to describe many-electron systems on the basis of electronic density instead of wave functions, although it was not very successful in the treatment of entire molecules. The method did provide the basis for what is now known as density functional theory. Though this method is less developed than post Hartree-Fock methods, its significantly lower computational requirements (scaling typically no worse than n with respect to n basis functions) allow it to tackle larger polyatomic molecules and even macromolecules. This computational affordability and often comparable accuracy to MP2 and CCSD (post-Hartree–Fock methods) has made it one of the most popular methods in computational chemistry at present.

Chemical Dynamics

A further step can consist of solving the Schrödinger equation with the total molecular Hamiltonian in order to study the motion of molecules. Direct solution of the Schrödinger equation is called quantum molecular dynamics, within the semiclassical approximation semiclassical molecular dynamics, and within the classical mechanics framework molecular dynamics (MD). Statistical approaches, using for example Monte Carlo methods, are also possible.

Adiabatic Chemical Dynamics

In adiabatic dynamics, interatomic interactions are represented by single scalar potentials called potential energy surfaces. This is the Born-Oppenheimer approximation introduced by Born and Oppenheimer in 1927. Pioneering applications of this in chemistry were performed by Rice and Ramsperger in 1927 and Kassel in 1928, and generalised into the RRKM theory in 1952 by Marcus who took the transition state theory developed by Eyring in 1935 into account. These methods enable simple estimates of unimolecular reaction rates from a few characteristics of the potential surface.

Non-adiabatic Chemical Dynamics

Non-adiabatic dynamics consists of taking the interaction between several coupled potential energy surface (corresponding to different electronic quantum states of the molecule). The coupling terms are called vibronic couplings. The pioneering work in this field was done by Stueckelberg, Landau, and Zener in the 1930s, in their work on what is now known as the Landau-Zener transition. Their formula allows the transition probability between two diabatic potential curves in the neighbourhood of an avoided crossing to be calculated.

Quantum Optics

Quantum optics is a field of research in physics, dealing with the application of quantum mechanics to phenomena involving light and its interactions with matter.

History of Quantum Optics

Light is made up of particles called photons and hence inherently is "grainy" (quantized). Quantum optics is the study of the nature and effects of light as quantized photons. The first indication that light might be quantized came from Max Planck in 1899 when he correctly modelled blackbody radiation. By assuming blackbody radiation is quantized, Bohr showed that the atoms were also quantized, in the sense that they could only emit discrete amounts of energy. The understanding of the interaction between light and matter following these developments not only formed the basis of quantum optics but was also crucial for the development of quantum mechanics as a whole. However, the subfields of quantum mechanics dealing with matter-light interaction were principally regarded as research into matter rather than into light; hence one rather spoke of atom physics and quantum electronics in 1960. Laser science—i.e., research into principles, design and application of these devices—became an important field, and the quantum mechanics underlying the laser's principles was studied now with more emphasis on the properties of light, and the name quantum optics became customary.

As laser science needed good theoretical foundations, and also because research into these soon proved very fruitful, interest in quantum optics rose. Following the work of Dirac in quantum field theory, George Sudarshan, Roy J. Glauber, and Leonard Mandel applied quantum theory to the electromagnetic field in the 1950s and 1960s to gain a more detailed understanding of photodetection and the statistics of light.

This led to the introduction of the coherent stateas a quantum description of laser light and the realisation that some states of light could not be described with classical waves. In 1977, Kimble et al. demonstrated the first source of light which required a quantum description: a single atom that emitted one photon at a time. This was the first conclusive evidence that light was made up of photons. Another quantum state of light with certain advantages over any classical state, squeezed light, was soon proposed. At the same time, development of short and ultrashort laser pulses—created by Q switching and modelocking techniques—opened the way to the study of unimaginably fast ("ultrafast") processes. Applications for solid state research (e.g. Raman spectroscopy) were found, and mechanical forces of light on matter were studied. The latter led to levitating and positioning clouds of atoms or even small biological samples in an optical trap or optical tweezers by laser beam. This, along with Doppler cooling was the crucial technology needed to achieve the celebrated Bose-Einstein condensation.

Other remarkable results are the demonstration of quantum entanglement, quantum teleportation, and (recently, in 1995) quantum logic gates. The latter are of much interest in quantum information theory, a subject which partly emerged from quantum optics, partly from theoretical computer science.

Today's fields of interest among quantum optics researchers include parametric down-conversion, parametric oscillation, even shorter (attosecond) light pulses, use of quantum optics for quantum information, manipulation of single atoms, Bose-Einstein condensates, their application, and how to manipulate them (a sub-field often called atom optics), coherent perfect absorbers, and much more.

Research into quantum optics that aims to bring photons into use for information transfer and computation is now often called photonics to emphasize the claim that photons and photonics will take the role that electrons and electronics now have.

Concepts of Quantum Optics

According to quantum theory, light may be considered not only as an electro-magnetic wave but also as a "stream" of particles called photons which travel with c, the vacuum speed of light. These particles should not be considered to be classical billiard balls, but as quantum mechanical particles described by a wavefunction spread over a finite region.

Each particle carries one quantum of energy equal to hf, where h is Planck's constant and f is the frequency of the light. The postulation of the quantization of light by Max Planck in 1899 and the discovery of the general validity of this idea in Albert Einstein's 1905 explanation of the photoelectric effect soon led physicists to realise the possibility of population inversion and the possibility of the laser.

This kind of use of statistical mechanics is the fundament of most concepts of quantum optics: Light is described in terms of field operators for creation and annihilation of photons—i.e. in the language of quantum electrodynamics.

A frequently encountered state of the light field is the coherent state as introduced by George Sudarshan in 1963. This state, which can be used to approximately describe the output of a single-frequency laser well above the laser threshold, exhibits Poissonian photon number statistics. Via certain nonlinear interactions, a coherent state can be transformed into a squeezed coherent state, which can exhibit super- or sub- Poissonean photon statistics.

Such light is called squeezed light. Other important quantum aspects are related to correlations of photon statistics between different beams. For example, parametric nonlinear processes can generate so-called twin beams, where ideally each photon of one beam is associated with a photon in the other beam.

Atoms are considered as quantum mechanical oscillators with a discrete energy spectrum with the transitions between the energy eigenstates being driven by the absorption or emission of light according to Einstein's theory with the oscillator strength depending on the quantum numbers of the states.

For solid state matter one uses the energy band models of solid state physics. This is important as understanding how light is detected (typically by a solid-state device that absorbs it) is crucial for understanding experiments.

Quantum Electronics

This term was used for the area of physics dealing with the effects of quantum mechanics on the behaviour of electrons in matter, and their interactions with photons. It is today rarely considered a sub-field in its own right, as it has been absorbed by other fields. Solid state physics regularly takes quantum mechanics into account, and is usually concerned with electrons. Specific application to electronics is researched within semiconductor physics. The term also encompassed the basic processes of laser operation where photons are interacting with electrons: absorption, spontaneous emission, and stimulated emission. The term was mainly used between the 1950s and the 1970s. Today, the research output of this field is mainly used in quantum optics, especially for the part of it that draws not from atomic physics but from solid-state physics. Its usage overlapped Quantum Hall effect and Quantum cellular automata.

Quantum Information Science

Quantum information science is an area of study based on the idea that information science depends on quantum effects in physics. It includes theoretical issues in computational models as well as more experimental topics in quantum physics including what can and cannot be done with quantum information. The term quantum information theory is sometimes used, but it fails to encompass experimental research in the area.

Subfields include:

- Quantum computing, which deals on the one hand with the question how and whether one can build a quantum computer and on the other hand, algorithms that harness its power
- Quantum complexity theory
- Quantum cryptography and its generalisation, quantum communication
- Quantum error correction
- Quantum communication complexity
- Quantum entanglement, as seen from an information-theoretic point of view
- Quantum dense coding
- Quantum teleportation is one well-known quantum information processing operation which reliably transfers an unknown quantum state from one point to another distant point, destroying the original state in the process.

Quantum Computer

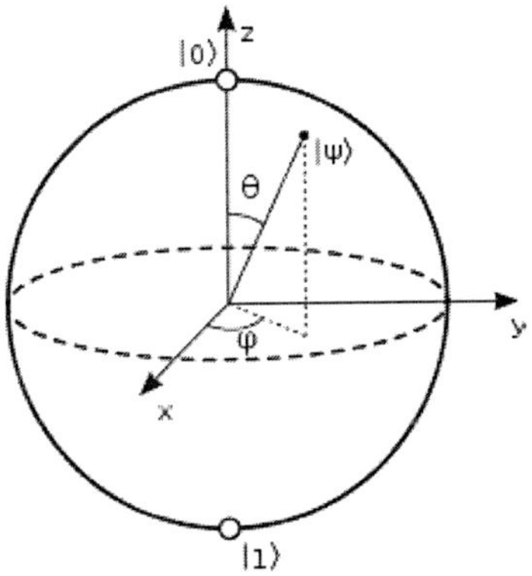

Figure: *The Bloch sphere is a representation of a qubit, the fundamental building block of quantum computers.*

A quantum computer is a device for computation that makes direct use of quantum mechanical phenomena, such as superposition and entanglement, to perform operations on data. Quantum computers are different from traditional computers based on transistors. The basic principle behind quantum computation is that quantum properties can be used to represent data and perform operations on those data. A theoretical model is the quantum Turing machine, also known as the universal quantum computer. Quantum computers share theoretical similarities with non-deterministic and probabilistic computers, like the ability to be in more than one state simultaneously. The field of quantum computing was first introduced by Richard Feynman in 1982.

Although quantum computing is still in its infancy, experiments have been carried out in which quantum computational operations were executed on a very small number of qubits (quantum bits). Both practical and theoretical research continues, and many national government and military funding agencies support quantum computing research to develop quantum computers for both civilian and national security purposes, such as cryptanalysis.

Large-scale quantum computers could be able to solve certain problems much faster than any classical computer by using the best currently known algorithms, like integer factorization using Shor's algorithm or the simulation of quantum many-body systems. There exist quantum algorithms, such as Simon's algorithm, which run faster than any possible probabilistic classical algorithm.

Given unlimited resources, a classical computer can simulate an arbitrary quantum algorithm so quantum computation does not violate the Church–Turing thesis. However, in practice infinite resources are never available and the computational basis of 500 qubits, for example, would already be too large to be represented on a classical computer because it would require 2 complex values to be stored. Nielsen and Chuang point out that "Trying to store all these complex numbers would not be possible on any conceivable classical computer."

Basis

A classical computer has a memory made up of bits, where each bit represents either a one or a zero. A quantum computer maintains a sequence of qubits. A single qubit can represent a one, a zero, or, crucially, any quantum superposition of these; moreover, a pair of qubits can be in any quantum superposition of 4 states, and three qubits in any superposition of 8.

In general a quantum computer with n qubits can be in an arbitrary superposition of up to 2 different states simultaneously (this compares to a normal computer that can only be in one of these 2 states at any one time). A quantum computer operates by manipulating those qubits with a fixed sequence of quantum logic gates. The sequence of gates to be applied is called a quantum algorithm.

An example of an implementation of qubits for a quantum computer could start with the use of particles with two spin states: "down" and "up" (typically written$|\downarrow\rangle$ and $|\uparrow\rangle$, or $|0\rangle$ and $|1\rangle$). But in fact any system possessing an observable quantity A which is conserved under time evolution and such that A has at least two discrete and sufficiently spaced consecutive eigenvalues, is a suitable candidate for implementing a qubit. This is true because any such system can be mapped onto an effective spin-1/2 system.

Bits vs. Qubits

A quantum computer with a given number of qubits is fundamentally different than a classical computer composed of the same number of classical bits. For example, to represent the state of an n-qubit system on a classical computer would require the storage of 2 complex coefficients. Although this fact may seem to indicate that qubits can hold exponentially more information than their classical counterparts, care must be taken not to overlook the fact that the qubits are only in a probabilistic superposition of all of their states. This means that when the final state of the qubits are measured, they will only be found in one of the possible configurations they were in before measurement. Moreover, it is incorrect to think of the qubits as only being in one particular state before measurement since the fact that they were in a superposition of states before the measurement was made directly affects the possible outcomes of the computation.

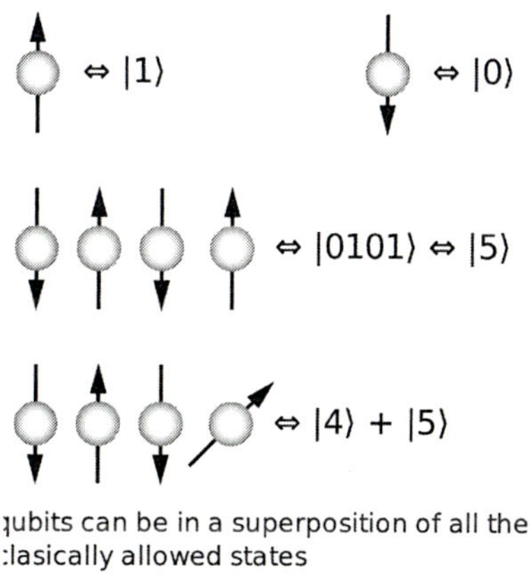

***Figure:** Qubits are made up of controlled particles and the means of control (e.g. devices that trap particles and switch them from one state to another).*

For example: Consider first a classical computer that operates on a three-bit register. The state of the computer at any time is a probability distribution over the 2 = 8 different three-bit strings 000, 001, 010, 011, 100, 101, 110, 111. If it is a deterministic computer, then it is in exactly one of these states with probability 1. However, if it is a probabilistic computer, then there is a possibility of it being in any one of a number of different states. We can describe this probabilistic state by eight nonnegative numbers a,b,c,d,e,f,g,h (where a = probability computer is in state 000, b = probability computer is in state 001, etc.). There is a restriction that these probabilities sum to 1.

The state of a three-qubit quantum computer is similarly described by an eight-dimensional vector (a,b,c,d,e,f,g,h), called a ket. However, instead of adding to one, the sum of the squares of the coefficient magnitudes, $|a|+|b|+...+|h|$, must equal one. Moreover, the coefficients are complex numbers. Since the probability amplitudes of the states are represented with complex numbers, the phase between any two states is a meaningful parameter, which is a key difference between quantum computing and probabilistic classical computing.

If you measure the three qubits, you will observe a three-bit string. The probability of measuring a given string is the squared magnitude of that string's coefficient (i.e., the probability of measuring 000 $=|a|$, the probability of measuring 001 $= |b|$, etc..). Thus, measuring a quantum state described by complex coefficients (a,b,...,h) gives the classical probability distribution $(|a|,|b|,...,|h|)$ and we say that the quantum state "collapses" to a classical state as a result of making the measurement.

Note that an eight-dimensional vector can be specified in many different ways depending on what basis is chosen for the space. The basis of bit strings (e.g., 000, 001, ..., 111) is known as the computational basis. Other possible bases are unit-length, orthogonal vectors and the eigenvectors of the Pauli-x operator. Ket notation is often used to make the choice of basis explicit. For example, the state (a,b,c,d,e,f,g,h) in the computational basis can be written as:

$$a|000\rangle + b|001\rangle + c|010\rangle + d|011\rangle$$
$$+e|100\rangle + f|101\rangle + g|110\rangle + h|111\rangle$$

where, e.g., $|010\rangle = (0,0,1,0,0,0,0,0)$

The computational basis for a single qubit (two dimensions) is $|0\rangle = (1,0)$ and $|1\rangle = (0,1)$.

Using the eigenvectors of the Pauli-x operator, a single qubit is $|+\rangle = \frac{1}{\sqrt{2}}(1,1)$ and $|-\rangle = \frac{1}{\sqrt{2}}(1,-1)$.

Operation

While a classical three-bit state and a quantum three-qubit state are both eight-dimensional vectors, they are manipulated quite differently for classical or quantum computation. For computing in either case, the system must be initialized, for example into the all-zeros string, $|000\rangle$, corresponding to the vector (1,0,0,0,0,0,0,0). In classical randomized computation, the system evolves according to the application of stochastic matrices, which preserve that the probabilities add up to one (i.e., preserve the L1 norm). In quantum computation, on the other hand, allowed operations are unitary matrices, which are effectively rotations (they preserve that the sum of the squares add up to one, the Euclidean or L2 norm). (Exactly what unitaries can be applied depend on the physics of the quantum device.) Consequently, since rotations can be undone by rotating backward, quantum computations are reversible. Technically, quantum operations can be probabilistic combinations of unitaries, so quantum computation really does generalise classical computation.

Finally, upon termination of the algorithm, the result needs to be read off. In the case of a classical computer, we sample from the probability distribution on the three-bit register to obtain one definite three-bit string, say 000.

Quantum mechanically, we measure the three-qubit state, which is equivalent to collapsing the quantum state down to a classical distribution (with the coefficients in the classical state being the squared magnitudes of the coefficients for the quantum state, as described above) followed by sampling from that distribution. Note that this destroys the original quantum state. Many algorithms will only give the correct answer with a certain probability, however by repeatedly initializing, running and measuring the quantum computer, the probability of getting the correct answer can be increased.

Potential

Integer factorization is believed to be computationally unfeasible with an ordinary computer for large integers if they are the product of few prime numbers (e.g., products of two 300-digit primes). By comparison, a quantum computer could efficiently solve this problem using Shor's algorithm to find its factors. This ability would allow a quantum computer to decrypt many of the cryptographic systems in use today, in the sense that there would be a polynomial time (in the number of digits of the integer) algorithm for solving the problem. In particular, most of the popular public key ciphers are based on the difficulty of factoring integers (or the related discrete logarithm problem which can also be solved by Shor's algorithm), including forms of RSA. These are used to protect secure Web pages, encrypted email, and many other types of data. Breaking these would have significant ramifications for electronic privacy and security.

However, other existing cryptographic algorithms do not appear to be broken by these algorithms. Some public-key algorithms are based on problems other than the integer factorization and discrete logarithm problems to which Shor's algorithm applies, like the McEliece cryptosystem based on a problem in coding theory. Lattice based cryptosystems are also not known to be broken by quantum computers, and finding a polynomial time algorithm for solving the dihedral hidden subgroup problem, which would break many lattice based cryptosystems, is a well-studied open problem.

It has been proven that applying Grover's algorithm to break a symmetric (secret key) algorithm by brute force requires roughly 2 invocations of the underlying cryptographic algorithm, compared with roughly 2 in the classical case, meaning that symmetric key lengths are effectively halved: AES-256 would have the same security against an attack using Grover's algorithm that AES-128 has against classical brute-force search. Quantum cryptography could potentially fulfill some of the functions of public key cryptography. Besides factorization and discrete logarithms, quantum algorithms offering a more than polynomial speedup over the best known classical algorithm have been found for several problems, including the simulation of quantum physical processes from chemistry and solid state physics, the approximation of Jones polynomials, and solving Pell's equation.

No mathematical proof has been found that shows that an equally fast classical algorithm cannot be discovered, although this is considered unlikely. For some problems, quantum computers offer a polynomial speedup. The most well-known example of this is quantum database search, which can be solved by Grover's algorithm using quadratically fewer queries to the database than are required by classical algorithms. In this case the advantage is provable. Several other examples of provable quantum speedups for query problems have subsequently been discovered, such as for finding collisions in two-to-one functions and evaluating NAND trees.

Consider a problem that has these four properties:

1. The only way to solve it is to guess answers repeatedly and check them,

2. The number of possible answers to check is the same as the number of inputs,
3. Every possible answer takes the same amount of time to check, and
4. There are no clues about which answers might be better: generating possibilities randomly is just as good as checking them in some special order.

An example of this is a password cracker that attempts to guess the password for an encrypted file (assuming that the password has a maximum possible length). For problems with all four properties, the time for a quantum computer to solve this will be proportional to the square root of the number of inputs. That can be a very large speedup, reducing some problems from years to seconds. It can be used to attack symmetric ciphers such as Triple DES and AES by attempting to guess the secret key.

Grover's algorithm can also be used to obtain a quadratic speed-up over a brute-force search for a class of problems known as NP-complete.

Since chemistry and nanotechnology rely on understanding quantum systems, and such systems are impossible to simulate in an efficient manner classically, many believe quantum simulation will be one of the most important applications of quantum computing.

There are a number of technical challenges in building a large-scale quantum computer, and thus far quantum computers have yet to solve a problem faster than a classical computer. David DiVincenzo, of IBM, listed the following requirements for a practical quantum computer:

- scalable physically to increase the number of qubits;
- qubits can be initialized to arbitrary values;
- quantum gates faster than decoherence time;
- universal gate set;
- qubits can be read easily.

Quantum Decoherence

One of the greatest challenges is controlling or removing quantum decoherence. This usually means isolating the system from its environment as interactions with the external world causes the system to decohere. This effect is irreversible, as it is non-unitary, and is usually something that should be highly controlled, if not avoided.

Decoherence times for candidate systems, in particular the transverse relaxation time T_2 (for NMR and MRI technology, also called the dephasing time), typically range between nanoseconds and seconds at low temperature.

These issues are more difficult for optical approaches as the timescales are orders of magnitude shorter and an often-cited approach to overcoming them is optical pulse shaping. Error rates are typically proportional to the ratio of operating time to decoherence time, hence any operation must be completed much more quickly than the decoherence time.

If the error rate is small enough, it is thought to be possible to use quantum error correction, which corrects errors due to decoherence, thereby allowing the total calculation time to be longer than the decoherence time. An often cited figure for required error rate in each gate is 10 . This implies that each gate must be able to perform its task in one 10,000th of the decoherence time of the system.

Meeting this scalability condition is possible for a wide range of systems. However, the use of error correction brings with it the cost of a greatly increased number of required qubits. The number required to factor integers using Shor's algorithm is still polynomial, and thought to be between L and L , where L is the number of bits in the number to be factored; error correction algorithms would inflate this figure by an additional factor of L. For a 1000-bit number, this implies a need for about 10 qubits without error correction. With error correction, the figure would rise to about 10 qubits. Note that computation time is about L or about 10 steps and on 1 MHz, about 10 seconds. A very different approach to the stability-decoherence problem is to create a topological quantum computer with anyons, quasi-particles used as threads and relying on braid theory to form stable logic gates.

Developments

There are a number of quantum computing models, distinguished by the basic elements in which the computation is decomposed. The four main models of practical importance are

- the quantum gate array (computation decomposed into sequence of few-qubit quantum gates),
- the one-way quantum computer (computation decomposed into sequence of one-qubit measurements applied to a highly entangled initial state (cluster state)),

- the adiabatic quantum computer (computation decomposed into a slow continuous transformation of an initial Hamiltonian into a final Hamiltonian, whose ground states contains the solution),
- and the topological quantum computer (computation decomposed into the braiding of anyons in a 2D lattice)

The Quantum Turing machine is theoretically important but direct implementation of this model is not pursued. All four models of computation have been shown to be equivalent to each other in the sense that each can simulate the other with no more than polynomial overhead.

For physically implementing a quantum computer, many different candidates are being pursued, among them (distinguished by the physical system used to realise the qubits):

- Superconductor-based quantum computers (including SQUID-based quantum computers) (qubit implemented by the state of small superconducting circuits (Josephson junctions))
- Trapped ion quantum computer (qubit implemented by the internal state of trapped ions)
- Optical lattices (qubit implemented by internal states of neutral atoms trapped in an optical lattice)
- electrically-defined or self-assembled quantum dots (e.g. the Loss-DiVincenzo quantum computer or) (qubit given by the spin states of an electron trapped in the quantum dot)
- Quantum dot charge based semiconductor quantum computer (qubit is the position of an electron inside a double quantum dot)
- Nuclear magnetic resonance on molecules in solution (liquid-state NMR) (qubit provided by nuclear spins within the dissolved molecule)
- Solid-state NMR Kane quantum computers (qubit realised by the nuclear spin state of phosphorusdonors in silicon)
- Electrons-on-helium quantum computers (qubit is the electron spin)
- Cavity quantum electrodynamics (CQED) (qubit provided by the internal state of atoms trapped in and coupled to high-finesse cavities)
- Molecular magnet
- Fullerene-based ESR quantum computer (qubit based on the electronic spin of atoms or molecules encased in fullerene structures)

- Optics-based quantum computer (Quantum optics) (qubits realised by appropriate states of different modes of the electromagnetic field, e.g.)
- Diamond-based quantum computer (qubit realised by the electronic or nuclear spin of Nitrogen-vacancy centres in diamond)
- Bose–Einstein condensate-based quantum computer
- Transistor-based quantum computer – string quantum computers with entrainment of positive holes using an electrostatic trap
- Rare-earth-metal-ion-doped inorganic crystal based quantum computers (qubit realised by the internal electronic state of dopants in optical fibers)

The large number of candidates demonstrates that the topic, in spite of rapid progress, is still in its infancy. But at the same time, there is also a vast amount of flexibility. In 2005, researchers at the University of Michigan built a semiconductor chip which functioned as anion trap. Such devices, produced by standard lithography techniques, may point the way to scalable quantum computing tools. An improved version was made in 2006. In 2009, researchers at Yale University created the first rudimentary solid-state quantum processor. The two-qubit superconducting chip was able to run elementary algorithms. Each of the two artificial atoms (or qubits) were made up of a billion aluminium atoms but they acted like a single one that could occupy two different energy states.

Figure: *Photograph of a chip constructed by D-Wave Systems Inc., mounted and wire-bonded in a sample holder. The D-Wave processor is designed to use 128 superconducting logic elements that exhibit controllable and tunable coupling to perform operations.*

Another team, working at the University of Bristol, also created a silicon-based quantum computing chip, based on quantum optics. The team was able to run Shor's algorithm on the chip. Further developments were made in 2010. Springer publishes a journal ("Quantum Information Processing") devoted to the subject.

A team of scientists from Australia and Japan have finally made a breakthrough in quantum teleportation. They have successfully transferred a complex set of quantum data with full transmission integrity achieved. Also the qubits being destroyed in one place but instantaneously resurrected in another, without affecting their superpositions.

In 2011, D-Wave Systems announced the first commercial quantum annealer on the market by the name D-Wave One. The company claims this system uses a 128 qubit processor chipset. On May 25, 2011 D-Wave announced that Lockheed Martin Corporation entered into an agreement to purchase a D-Wave One system .

Relation to Computational Complexity Theory

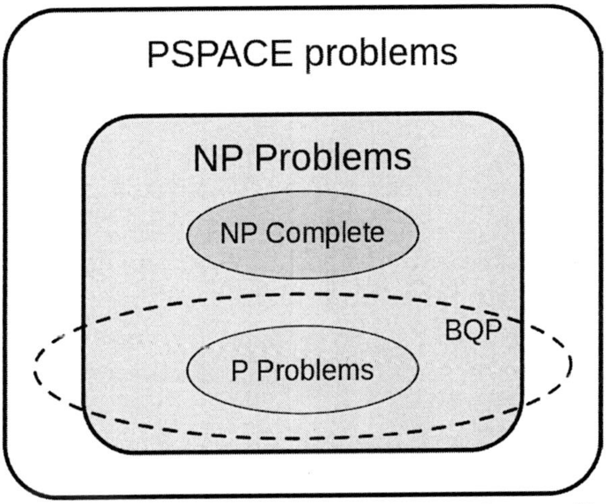

***Figure:** The suspected relationship of BQP to other problem spaces.*

The class of problems that can be efficiently solved by quantum computers is called BQP, for "bounded error, quantum, polynomial time". Quantum computers only runprobabilistic algorithms, so BQP on quantum computers is the counterpart of BPP ("bounded error, probabilistic, polynomial time") on classical computers. It is defined as the set of problems solvable with a polynomial-time algorithm, whose probability of error is bounded away from one half. A quantum computer is said to "solve" a problem if, for every instance, its answer will be right with high probability. If that solution runs in polynomial time, then that problem is in BQP.

BQP is contained in the complexity class #P (or more precisely in the associated class of decision problemsP), which is a subclass of PSPACE.

BQP is suspected to be disjoint from NP-complete and a strict superset of P, but that is not known. Both integer factorization and discrete log are in BQP. Both of these problems are NP problems suspected to be outside BPP, and hence outside P. Both are suspected to not be NP-complete. There is a common misconception that quantum computers can solve NP-complete problems in polynomial time. That is not known to be true, and is generally suspected to be false.

Possibilities of the quantum computer to accelerate classical algorithms has rigid limits — upper bounds of quantum computation's complexity. The overwhelming part of classical calculations cannot be accelerated on the quantum computer. A similar fact takes place for particular computational tasks, like the search problem, for which Grover's algorithm is optimal.

Although quantum computers may be faster than classical computers, those described above can't solve any problems that classical computers can't solve, given enough time and memory (however, those amounts might be practically infeasible). A Turing machine can simulate these quantum computers, so such a quantum computer could never solve an undecidable problem like the halting problem. The existence of "standard" quantum computers does not disprove the Church–Turing thesis. It has been speculated that theories of quantum gravity, such as M-theory or loop quantum gravity, may allow even faster computers to be built. Currently, it's an open problem to even define computation in such theories due to the problem of time, i.e. there's no obvious way to describe what it means for an observer to submit input to a computer and later receive output.

Quantum Complexity Theory

Quantum complexity theory is a part of computational complexity theory in theoretical computer science. It studies complexity classes defined using quantum computers and quantum information which are computational models based on quantum mechanics. It studies the hardness of problems in relation to these complexity classes, and the relationship between quantum complexity classes and classical (i.e., non-quantum) complexity classes.

A complexity class is a collection of problems which can be solved by some computational model under resource constraints. For instance, the complexity class P is defined to be the set of problems solvable by a Turing machine in polynomial time. Similarly, one may define a quantum complexity class using a quantum model of computation, such as a standard quantum computer or a quantum Turing machine. Thus, the complexity class BQP is defined to be the set of problems solvable by a quantum computer in polynomial time with bounded error.

Two important quantum complexity classes are BQP and QMA which are the bounded-error quantum analogues of P and NP. One of the main aims of quantum complexity theory is to find out where these classes lie with respect to classical complexity classes such as P, NP, PP, PSPACE and other complexity classes.

Chapter 7

Quantum Theory

Wave Mechanics

We have traversed the realms of physical research as rapid travelers. Since we failed to take along the heavy weapons of mathematics, we have been limited, so to speak, to the role of a civilian war correspondent in the land of physical research who must be content to draw some mood pictures of the events there without seriously pursuing the strategical and technical problems.

We passed by rich departments of knowledge, varied and beautiful in their content. To omit consideration of the thought structures in which modern physical thought triumphs over the apparently hopeless paradoxes and difficulties which were evidenced in quantum physical experiments would require still greater decision. It is not possible to relate and explain here the heavy weight of experimental quantitative proof of the content and mathematical laws of modern quantum theory; a clarification of the philosophical-logical character of this theory by intimation must suffice. But even for that the author, who wants to make it more easily accessible to the reader, must request special attention.

Let us consider briefly the historical development of modern quantum theory. The investigation of atomic spectra (and the energy levels related to them) presented physical research with an abundance of varied problems, in the step by step clarification of which physicists gradually worked out an understanding of the regularities of quantum physics. Niels Bohr, who started this development in 1913 also remained its leader. Without going into the thousand different problems of this field we shall consider the Bohr "correspondence principle". Although it was mentioned in reference to the energy or relativity principle —

and not opposing these in its meaning — the correspondence principle has a quite different character from the energy and relativity principles.

These latter are natural laws in finished, perfected form; their content can be expressed clearly and thoroughly in a few words. But the correspondence principle presents peculiar difficulties to the intelligence since its general content can only be described by intimation, in special individual examples. For the correspondence principle is not a finished, clearly definable law of nature, but indicates the direction in which Bohr's conviction about the solution of the quantum puzzle was to be sought, and in which he wished to steer the reflection of the quantum investigators.

It only becomes clear in the history of quantum theory that not only individual great discoveries are decisive for the development of our knowledge but that under certain circumstances the change of our spiritual attitude towards problems can be much more important. Physical thought develops not only from combining the compilations of results of individual observations and their logically-exact treatment; but rather the attainment of decisive new intelligence depends essentially on a creative phantasy, which for its part depends essentially on spiritual hypotheses and on the attitude our intuition assumes towards these things.

Bohr's historical contribution lies not only in the individual, pioneer discoveries through which he enriched quantum theory but also in the penetrating force his spirit exerted for the creation of a new spiritual "atmosphere" — wherein the problems were first gradually elucidated as the essential became distinguishable from the unessential and effective control of the solution became possible. The meaning of the correspondence principle is not purely historical; it is still indispensable today in the sense that it teaches us to view modern quantum theory knowledge with the proper attitude. The differences between quantum theory and classical physics are so deeply rooted that many physicists were tempted to regard the ideas of classical physics as absolutely useless for the comprehension of atomic physics; they attempted to introduce arbitrarily invented, new, untraditional ideas. But Bohr had — and this is the content of the correspondence principle — nergetically pointed out that in all individual problems a verifiable close similarity exists between classical and quantum theory despite their fundamental difference.

The following is a rough example. classical physics predicts that atoms, which con sist of electrically charged particles, necessarily emit light in the execution of internal motions. This prediction is correct. The elementary processes of this light emission do proceed quite differently from macroscopic antennae; but the fact remains that the fundamental expectation is fulfilled.

That is an example of the close relationship which exists between classical and quantum theory despite their incisive difference. There are other examples of this relationship, much finer examples; and through Bohr we gradually learned to see that such kindred relationships can absolutely be uncovered for each concrete individual problem through more exact analysis. In his emphasis on the affinity of classical and quantum theory there exists, however, a decided prominence of the independence of the quantum theory from classical theory.

Since we learn to "understand" various quantum-physical individual problems better through uncovering kindred relationships "in the manner of correspondence", we gradually attain one of the concepts of quantum physics used in classical theory but separated from it by fundamental differences and independent in itself.

Such a discovery is a radical challenge to the repeated spasmodic attempts throughout the course of historical development to somehow want to "explain" the characteristic quantum phenomena — to want to reduce them to ideas which conform more closely to the classical. With the progress of this development it gradually became clear which problems in general should be answered by a logically effected quantum theory. Here the positivist consideration must be repeated very forcefully; the goal of a purely intellectual comprehension of quantum phenomena must be a description of the experimental facts themselves. The experimental facts involved here are collectively of the following sort: a specific atom, characterized by its atomic number, according to experience possesses quite definite energy levels.

The first problem appears; how can the positions of all these energy levels (for any atom, and likewise for molecules) be determined and calculated from general theoretical principles? Furthermore, we find "transition-probabilities" for these atoms. When an atom exists in an energy-rich condition, after some time it will undergo a quantum transition which brings it down to a lower energy level while the energy freed thereby comes off as a light quantum.

The atom has a range of various such possibilities since there are different lower energy levels at its disposal. As in the case of the radium atom described above, it is not possible to predict in an individual case when and where the atom will jump. But experience indicates the existence of quite definite probabilities for the various possible processes. If the atom is irradiated with light there results a definite probability for a quantum transition associated with the absorption of a light quantum.

If an electron of a certain (sufficiently high) velocity impinges on an atom, the atom can, as we already know, thus also be induced to a quantum transition. If, furthermore, the atom collides with another atom (of the same or a different kind) an energy change can occur, as was already mentioned, in the form of a simultaneous mutual quantum transition. Definite probabilities prevail for all these processes; based on experience they are always stated in the same ways through the conditions of the experiment in question.

They are summarized under the designation of "transition probabilities". Now the problem (besides the theoretical determination of energy levels) that a complete quantum theory must solve quite comprehensively is simply the theoretical determination of the transition probabilities. In this sense Heisenberg undertook the creation of a "quantum mechanics". He relied upon the success then already achieved in the detailed execution of Bohr's correspondence principle; in fact, the systematic evaluation of Heisenberg's extensions (Heisenberg, Born, Dirac, Jordan) yielded a solution of the problem formulated according to the correspondence principle which was complete in principle.

Of course the mathematical formulation of this solution was such that it differed extensively from the mathematically precise renditions of our classical physical conceptions. But this mathematical form of the new quantum mechanics was most exactly suited to the problem (as we stated it) of the determination of transition probabilities.

Besides, this new quantum mechanics expressed wonderfully the two-sided relation of quantum to classical theory; namely, fundamental difference on the one side and close connection on the other.

These investigations had just reached a preliminary rounding-off point when Schrödinger arrived at the same goal by an entirely different method. Schrödinger started with de Broglie's investigations.

After de Broglie had shown how the uniform rectilinear motion of a corpuscular particle was reinterpreted in the wave theory, Schrödinger wondered how these de Broglie considerations developed for motions influenced by external forces.

For this reason he investigated the motion of an electron under the influence of the attraction of a positive heavy nucleus; and thus he arrived at the quantitative description of the electron charge cloud in the hydrogen atom.

Schrödinger could now show that with the solution of the mathematical problem he formulated he had simultaneously also achieved the solution for the apparently quite different mathematical problem which was expressed by quantum mechanics in the form of matrix theory.

Thus, when a problem has been solved by means of Schrödinger's "wave mechanics" a mathematical conversion yields the "quantum mechanics" solution for this same problem. This mathematical connection between the two theories, leading to the same result, which in view of the complete difference of the two methods must appear very surprising at first, also supplied the certain basis for the physical-abstract interpretation of Schrödinger waves.

At first it might appear that the discovery of wave mechanics yielded a mitigation of the contrast between quantum and classical theory; for here instead of the unclassical concepts of transition probabilities, etc., we deal with waves — something closer to classical ideas.

But what we just learned with respect to light is still true; the classical wave theory is not the final word. Wave mechanics does not in any way signify for atomic physics the removal or mitigation of the fundamental unclassical characteristics of quantum physics.

Also, for atoms with more than one electron (thus for all atoms except hydrogen) Schrödinger wave mechanics assumes a very abstract form; in these cases Schrödinger waves are no longer waves in customary threedimensional space but are simply a mathematical construction which mathematicians can "illustrate" to themselves as waves in space of more than three dimensions.

This abstract, multidimensional space can be avoided through another (mathematically equivalent) method of representation, the construction of which was a special hobby of the author's. In this method of representation ("second quantization"), which clings especially closely to the fundamental dualism of waves and corpuscles, the waves dealt with are spread out in ordinary three-dimensional space, but can only be described by means of the ideas of quantum mechanics.

These theories, whose abstract nature will not remain hidden from the reader in even this fleeting explanation, made possible a number of special uses for individual problems of atomic physics. The precise correctness of the new theory was confirmed without exception in a tremendous field of experimental results. No end is in sight for the further elaboration of special problems on the basis of these principles.

The fundamental result may be stated as follows: today we can understand all phenomena which occur near and in the electron shells. Thus all the elementary processes that are the basis for spectral or chemical processes are defined as clearly as the motions of the planetary systems have been since Newton. The only partially explored realm at present remains the physics of internal events in atomic nuclei.

For us it is only important to understand the philosophical nuclei of these new ideas. We again owe special thanks to Bohr and Heisenberg for the philosophical-epistemological explanation of the meaning and significance of the new theories. This became possible through the earlier thorough mathematical understanding of the quantum laws in the so-called "statistical transformation theory".

Objectivity

We return once again to classical theory to emphasize still more clearly the characteristic features through which it differs from quantum theory. We can state three basic principles which should be considered as the most important characteristics of classical theory with the catchwords continuity, causality, objectivity.

We spoke amply of continuity; and we learned that just this principle of continuity constitutes a difference between classical and quantum theory. But now we want to make it clear that continuity is not only something familiar to us, but is really an essential, the omission of which is followed with logical necessity by further fundamental deviations from classical ideas.

The assumed continuity of natural processes is essential for our method of executing and evaluating physical measurements. Every measurement we perform is inexact. The physicist considers the "limit of error" for each measurement; for every measured numerical value he records two numbers, the difference between them corresponding to the uncertainty of the measurement.

That it is nevertheless possible to draw certain conclusions from an inexact measurement — conclusions, of course, that are also affected by an uncertainty but in any case are rich in content and very definite — is only because the principle of continuity is actually fulfilled in the macrophysical world; trivial changes in the cause are followed by trivial changes in the effect.

Consequently an inexact knowledge of causes is still sufficient for an inexact prediction of the effects, although the absolutely certain prediction of the effects becomes possible only with a mathematically more precise (practically unattainable) measurement of the causes.

Uncertainty in the process of measurement deserves a still more detailed investigation. The essential point for us here will become clear if we visualize, for example, the measurement of the temperature of a macrophysical body. We bring the body in contact with a thermometer; the thermometer assumes the same temperature as the body in question; and we read off the value.

But to be exact, it is necessary to consider that since the body under investigation imparts some of its heat to the thermometer the body itself is influenced and its original condition is altered somewhat.

To avoid resultant false measurements the thermometer selected must be much smaller than the body being investigated. (Of course it is possible, if the thermometer used was not sufficiently small, to compute or estimate the temperature change which took place and correct the result for it; but that is a problem in technical method which has no connection with the nature of the thing.)

It is analogous for every physical measurement: I must always select such fine instruments for a measurement that in the measuring process the body under investigation itself will not be influenced appreciably by too rough a measuring instrument (such influence would falsify the measured result).

Fundamentally a reaction of the measuring instrument on the object being investigated is inherent in every physical measurement; yet in the investigation of macrophysical objects this reaction can be made sufficiently small by the selection of quite fine instruments.

This idealization of the process of measuring in classical physics depends essentially on the assumption of continuity of all natural processes in the sense that our fundamental considerations are established as though precise measurements were possible for us. It is a "permissible idealization" to base our considerations on this concept of absolute observational accuracy and to interpret the actual inaccuracy of each measurement as an only practically and not principally significant secondary disturbance.

This idea cannot lead us to errors, because by classical concepts we can approach ideal measurement unlimitedly, although we can never attain it. We here consider something which we shall indicate with the word objectivity and which is so characteristic of our general classical-physical thinking that we usually disregard it. We are accustomed to regarding physical observation results as clearly understood according to their importance if we have explained them as effects of an objective physical process or condition.

This formulation hides within it a higher type of positivist modesty; a deeper "explanation" of natural processes is relinquished herein for everything. But what is left after this renunciation of all classical-physical theories as a basis for their methods of representation is just this idea in objective events.

Perhaps we do not describe planetary motions by specifying when and where or through what telescope the various planets were observed; but we do describe planetary motions as a spatial-temporal process taking place independently of human observation.

Naturally one can pursue the familiar philosophical considerations which emphasize that without subjects doing the observing there can be no talk of objective matters. But these thought processes are not acceptable to the physicist.

To him the motion of the planet Neptune in its orbit is an objective event which was already in process before anyone had seen this planet in a telescope and which continues uninfluenced, independent of whether or when or how often it is observed or photographed. Exactly the same self-evident supposition of objective events is the basis for Maxwellian electrodynamics; we interpret electrical measurement data as indication of an objective physical event in the electromagnetic field which is present in space.

Positivist criticism must remind us that this objectivity of physical events is not a purely logical self-evident fact. For positivism teaches us to view true physical reality only in the totality of experimental results. It is very remarkable and astonishing that in the domain of validity of macrophysics we are in a position to so formulate our summarizing description of experimental results that they are no longer referred to directly but are, so to speak, only minor appendages of the picture we have traced out — a picture which maintains the existence of objective events which occur independently of how and where the observations necessary for their detection are taken.

After what has been said it is obvious that the idea of ideal measurements (made possible by the continuity of macrophysical processes) which do not disturb the observed event in the least is in dispensable for this construction of an objective physical world. Another assumption indispensable for this objectivity is that of complete causality in the macrophysical world.

For if we had had to reckon with the occurrences of effects not determined strictly causally we could never have been certain in our observation processes whether the source of an effect we saw was to be sought in the object we were observing or just in a cause-free reaction of our observation instrument.

Thus objectivity would be disturbed if complete causality did not exist; conversely, the representation of objective physical events is a necessary assumption for the carrying through of the idea of strict causality. That complete causality is an indispensable assumption for the possibility of effecting the representation of objective physical events was already clearly recognized by Kant.

But there is no justification for concluding from this insight that the complete validity of the principle of causality throughout nature is guaranteed from the outset independently of experimental experience. All that must be established is this — objectivity also is weakened with the renunciation of complete causality. This is actually the case in atomic physics; we have seen that the principle of continuity can not be given up without objectivity ceasing.

And we have learned that continuity ends in atomic and quantum physics. We can consider the appearance of discontinuities in elementary physical processes as the fundamental proposition of modern quantum physics; the atomistic structure of matter may be in terpreted as one facet of this elementary physical discontinuity which also appears in many other forms.

The above examples have already made it clear that in quantum physics a complete causality, of the type we are accustomed to, no longer exists. But we still want to refer particularly to the difference that exists between the quantum theory use of the probability concept and the Boltzmann evaluation of it (for the kinetic explanation of heat processes).

In Boltzmann's considerations statistics was a secondary thing; at that time there was no reason to doubt that the motion of each individual atom could fundamentally be calculated precisely in advance.

The pursuit of such fine processes was voluntarily relinquished and statistical considerations were used as an expression of an incomplete (but sufficient for the result desired) observation of the events. Whereas in quantum theory the primary natural laws themselves take the form of probability expressions in this case the statistical concepts are not an expression of the incompleteness of our insight into events, but rather an expression of an indefiniteness existing in nature itself.

Nature herself did not determine individual atomic processes in advance; from case to case she executes unpredictable decisions which show fixed regularities only in the statistical average. But these unpredictable decisions of nature are always connected with elementary quantum physics discontinuities. Indeed it is the independent cases of quantum transitions that are not predetermined.

Our earlier example of the interference of light for the screen with two openings already demonstrated that quantum physics must also relinquish the idea of objective events. With characteristic positivist modesty we limited our problem to the problem of how great the probability is that the light quantum in question will be absorbed in a certain point of the photographic plate.

Thus two quantum transitions — and that is typical of the whole modern quantum theory -are placed in a static relation with each other. We observe the quantum transformation of the light emission from the point source; then we observe the quantum transformation of the absorption of the light quantum in a certain grain of the plate; and we can theoretically calculate in advance the probability of the occurrence of the second elementary act after the incidence of the first. But we cannot extend the picture of an objective event between both processes in the form of the specification of a continuous path which the light quantum must traverse from the first position to the second.

Formerly one might have been inclined to suspect that the principle of the objectivity of physical events was also inseparably connected with the possibility of the quantitative, mathematical comprehension of natural regularities. But we see that that is plainly incorrect; everything we try to explain here in words can be expressed in as mathematically clear a form as Galileo-Newton mechanics. We must replace the precalculation of future events on the basis of complete causality possible in the macrophysical domain by a purely proba bility prediction. The probabilities of quantum physics processes are themselves determined exactly quantitatively; in the last analysis they are subject to precise mathematical laws of great simplicity and of the most comprehensive validity.

Despite that, the Kantian interpretation, that complete causality (as well as objectivity and continuity) of natural processes is an indispensable provision of every physical thought in general, is still correct in the following sense: quantum physics experiments are also always performed with macrophysical apparatus. We need macrophysical (thus functioning according to strict causality) apparatus to be able to make any correct observations at all and to be able to determine regularities in the atomic world. Classical physics remains the indispensable support from which an advance into the .world of quanta and atoms becomes possible.

This can not be altered by the fact that macrophysical laws can naturally be interpreted as results of quantum physics elementary laws. The laws which govern the motions of macrophysical bodies must naturally result from the laws to which their individual atoms are subject.

There is no difficulty involved in the necessity of interpreting the strict causality of macrophysical events as a result of the purely statistical laws for the elementary processes. For that is the essence of conformity to statistical laws — despite the incalculability of the separate event, an exact, predictable result occurs in the total effect of a large number of individual processes. The fact that we thus interpret the laws of atomic physics as the really true natural laws from which macrophysical laws are derived as results does not permit us to overlook the other fact, that the elementary laws of atomic physics include a tangible content only when they are attached to the frame of macrophysics by concrete application. Herein lies in the final analysis the root of the not only historical, but contemporary significance of the Bohr correspondence principle. For this teaches us to understand the meaning and content of quantum physics laws from their relation to macrophysics.

Complementarity

Experimental evidence has shown us in a most comprehensive way and with a variety exceeding all expectations the atomistic structure of all physical substrata; not only matter, but also light (despite its wave nature which it exhibits "on the other side") has a corpuscular make-up. This evidence drives us to remarkable conclusions.

How can one observe and investigate individual atoms? We saw that modern experimental technique permits the very satisfactory performance of experiments immediately involved with individual atoms (and therewith to exactly prove the reality of these atoms, without a doubt). Naturally experimental manipulation of individual atoms remains much more difficult and far different fromthe investigation and measuring of macrophysical, visibly large bodies which are made up of innumerable atoms. For macrophysical bodies we have scales and other mechanical, optical or electrical measuring instruments at our disposal; but there is a decisive difference for measurements on atoms. We know that every physical substratum, therefore every physical measuring instrument, is composed of atoms — be they material atoms or electrons or light quanta. This destroys all possibilities of using convenient measuring instruments for this research as is done for macrophysical objects or events.

We have already considered how fundamental for our classical physical ideas and methods of perception is the fact that in macrophysical investigations the back-coupling, the influencing of the object by the observation process, can be made negligibly small through the use of sufficiently fine measuring instruments. But if we consider that all measuring instruments themselves consist of atoms (thus can never be made finer and smaller than single atoms) we see that this method of eliminating the reaction of the measuring instrument is barred when the object to be investigated is itself an individual atom (or a structure containing just a few atoms).

There is no longer any possibility of investigating and observing with instruments that are finer than the object in question. Nor is it possible to control the influencing of the object by the measuring instrument and to eliminate it from the result by a corresponding conversion. Thus it must be considered as part of the bargain that fundamentally measurments on atomic objects are always "falsified" in the sense that according to natural law the object experiences a variable interference from the execution of the observation process.

Similarly when we try to observe and psychologically control our own thought processes we are successful up to a certain degree; but the functioning of the observation itself again influences the observed "object" — our own thought process — and with this sets the limits for the possibility of observation. (One can not, e.g, watch through psychological self-observation how one falls asleep, because just that attention of observing prevents one from falling asleep.

Also, e.g., the origin of voluntary decisions is disturbed and changed by internal-controlling self-observation.) Similarities are also found in atomic physics — the observed object is influenced by the observation process itself. With Niels Bohr we can say, the separation between observed object and observing subject begins to vanish here.

It is not at all proper to call this influencing of the object by the process of observation a "falsification". For we are not dealing with a disturbing influence which is in any way limited for the present by the current deficiency of our observation technique; these barriers to the possibility of an ideal observation which does not influence the object itself are limited — through the atomistic structure of physical foundations — by natural law.

Therefore, we must ascribe to the atomic objects themselves a certain character of "indeterminateness", of "indefiniteness" of their physical behaviour which makes the construction of an objective picture of atomic physical events impossible. This need not imply that every possibility of exact measurement on atomic objects is doomed to failure. It is possible throughout to make each physical property of an atom the object of precise measurements.

If a certain property of an atom is observed exactly there result from the observation process (due to the reaction of the measuring instrument to the object) strong uncontrollable and undefined changes (strong "uncertainties") with regard to other properties of this atom.

By optionally transferring the interference which is necessarily associated with observation of the atom to different properties of the object it is possible to make accurate observations on the properties of the atom undisturbed by the particular reaction. With this "complementarity", as Bohr named it, the new theory can eliminate those apparently hopeless contradictions we first encountered in the dualism of waves and corpuscles and which we met at every step in quantum physics.

This idea of complementarity must be viewed as the most significant result for philosophy that crystallized out of modern physics. It presents an absolutely new scientific way of thinking which is fundamentally different from classical scientific thinking in terms of objectivized representations. After the intellectual penetration and comprehension of atomic physical phenomena completely inaccessible to the previous method of representation became possible through it, it appears justified to believe that it may further become of epoch-making importance in other realms of natural science.

The enlightening force of this idea in solving an apparently insoluble puzzle and contradiction is clearly demonstrated in the famous problem of the dual istic nature of light. The properties connected with the wave nature of light on the one hand and those connected with its corpuscular nature on the other are "complementary" to each other in the sense that they can never appear in one and the same experiment at the same time (and thus come into actual direct opposition).

Experiments which let the wave side of light emerge clearly force (through the action that is connected with every experiment) the corpuscular nature of light back into the indeterminate and unobservable; other experiments, which force the corpuscular side of light into prominence, leave undefined and indiscernible all the properties which usually betray to us the wave nature of light. With this wonderful device of complementarity nature combines in one and the same physical object properties and regularities that contradict each other so that they could never exist directly at the same time.

Let us pursue this in somewhat more detail in the already repeatedly discussed example of the interference of light rays transmitted through two screen openings. If we want to let the interference of these two light rays take place to the wave nature of light we must, as already indicated, relinquish the desire to determine simultaneously through which of the two openings a definite light quantum was transmitted.

We can also consider this event from the "complementary" opposite side. We can undertake to desire to observe through which opening the light quantum was transmitted. But then we must conversely relinquish any desire to obtain an interference of the two rays. For with the latter intention the only procedure we can follow is to roughly close up one of the screen openings in order to be certain that a transmitted light quantum could actually only have passed through the other opening.

There is no experimental possibility of assuring ourselves in any way through which of the openings the light quantum was transmitted without simultaneously so altering the conditions of the experiment that the interference of the two rays is hindered.

As another example let us consider the charge cloud of the electron in the hydrogen atom. If we fix the position of a definite point with a microscope we must assume an inaccuracy which is at least of the order of magnitude of the wave lengths of visible light — thus much larger than an atom. But nothing hinders our imagining that we have a microscope that is not dependent on visible light, but on X-rays or even rays of much shorter wave length (gamma rays). With this microscope we could accomplish measurements of position on an electron with such great accuracy that positions inside of the charge cloud of the hydrogen atom would also be distinguished.

Now let us place a hydrogen atom under this gamma-microscope and "examine" the inner structure of this atom so that we determine the position of the electron. In the description of this imaginary experiment we must not forget one thing; namely, that the energy of the "gammalight" necessary to illuminate and render the electron visible is concentrated in individual light quanta which possess very high energy because of the minuteness of their wave lengths. The procedure of the "examination" of the electron in the hydrogen atom is as follows: a single energyrich light quantum meets the electron, and reflected by it approaches us through the microscope, showing us the exact position of the electron.

But in this process the electron experiences a terrific effect; it is thoroughly dislodged from its former condition by the interaction with the energy-rich light quantum and in most cases we can expect that the electron is completely torn away from the hydrogen nucleus; the atom, thus, is ionised. We can now explain in a concrete way the meaning of the "charge cloud" around the atomic nucleus calculated according to wave mechanics. We repeat the above described experiment innumerable times; and, it must be emphasized, each time we use a hydrogen atom which is in its lowest energy state (normal state). If, instead, we took atoms in a fixed, higher energy level we would have to reckon with a different structure of the charge cloud; for each of the different energy levels, according to quantum theory, there exists a certain specific structure of the charge cloud.

By repeatedly measuring the position of the electron in a hydrogen atom in its normal state we find the electron in different positions from case to case, just as in our interference experiment we found the individual light quanta in different places on the photographic plate, statistically distributed according to the light intensity calculated by the wave theory. The statistical distribution of the individually measured electron positions is given by the charge cloud calculated according to wave mechanics. It was only with this determination that the concept of this charge cloud attained a clear meaning which is defined by concrete, demonstrable experiments.

At the same time we see in this experiment how the changing actions in quantum physical observation are bound up with the observed facts. Before the act of observation in question the hydrogen atom possesses a definite energy, but in general in these circumstances a definite position of the electron does not exist — the position of the electron is undefined, or only indefinitely defined in the structure of the statistical charge cloud. It is only through the process of observing its location in the gamma-ray microscope that we force the electron to assume a definite position.

Notice, we do not prescribe in which position it should emerge; but we do force it into some definite position and thereby force it to a new crisis; now the electron assumes a definite position but simultaneously an indefinite energy exchange has taken place between the gamma light quantum and the electron and the original condition of a defined energy of the atom has been disturbed. A quite analogous process is represented by the previously described impinging of a linearly polarized light quantum against a Nicol prism. In this case we can also say that the execution of an act of observation has forced the light quantum to assume a clearly defined situation, when previously this was indefinite.

The light quantum must decide whether to be transmitted by the Nicol prism which has been placed inclined to the original plane of vibration of the light quantum or to be reflected by it. That is a decision of the same kind as that of assuming a definite position forced upon the electron in the above experiment. It is obvious that there is no more place among these ideas for a complete causality, clearly predetermining each occurrence. If in a macrophysical structure, the planetary system, for example, we want to pre-calculate future movements exactly we must know two things. First, we must know that the Newtonian law is valid (and not any other one), and we must know the magnitudes of the different planets which are determinative for Newtonian gravitational attraction and for the relations of force and acceleration.

Secondly, for any one point in time we must know the positions and velocities the different planets possess at exactly this point in time. Thereby the general course of motion is mathematically precisely determined for all later (besides, also earlier) times. In an electron, however, we are not able to simultaneously determine its position and velocity at a definite point in time.

Since position and velocity are complementary, the position measurement in the gamma-microscope makes the velocity of the electron unobservable — and the converse also holds. After knowing that the physical properties of an atom are partially complementary to one another, that therefore it is impossible to observe the atom simultaneously "from all sides" as it were (as is possible in macrophysical bodies), it must be regarded as quite natural that precalculations of the future conduct of atoms, electrons and light quanta are not possible analogously as in the planets. Let us emphasize once again: this impossibility not only depends on a practical technical incompleteness of our instruments, but depends on nature itself. It is a positive result of the natural laws which in quantum or wave mechanics have attained a formulation which is mathematically exact and is verified by innumerable experiments.

The indefiniteness inherent in the physical condition of all atomic objects stipulates a corresponding indefiniteness in the process of the action; pre-calculation according to exact causal laws is lacking here. The inability of the physicist to predict for an individual case which of the various possibilities will be realised in the quantum transition which is the basis for an observation process is not due to human imperfection of knowledge; nature herself has reserved until the last the decision for each individual case.

Chapter 8

Magnetic Field

Intramolecular Forces

The electrical force which is responsible for all attractive interactions between atoms, molecules, and ions. The electrical force is described quantitatively by Coulomb's Law. A system in which forces act can possess two types of energy. These are kinetic energy, KE, and potential energy, PE. The magnitude of the kinetic energy is a function only of temperature, as we have seen. Potential energy arises from the action of the Coulomb force.

We will discuss the various ways in which Coulomb forces operate in atomic/molecular systems, manifesting in both intramolecular and intermolecular contexts. Intramolecular forces include the forces between atoms in covalent molecules; the forces between oppositely charged ions in ionic compounds; and the forces between identical atoms in metals. Intermolecular forces include the forces that operate between one atom or molecule and another, or between the ion of one substance and the molecules of another. Although there are differences among the various types of forces, they are all fundamentally similar: they give rise to potential energy between the interacting particles, described by the potential well.

We shall see that an interplay between the depth of this well and the thermal kinetic energies of particles is responsible for many of the physical properties of substances.

Types of Energy and the Coulomb Force

An atom or a collection of atoms can possess energy of only two types. Kinetic energy (abbreviated KE) is energy of motion. A body of mass, m, moving with velocity, v, has kinetic energy given by equation.

$$KE = mv^2/2$$

When mass is expressed in kg and velocity in m/s, KE has units of kg m^2/s^2, orJoules (J). One Joule is defined as 1 kg m^2/s^2. The kinetic energy of atoms and molecules depends only on temperature, through the simple expression in 6-1-2.

$$\text{Average KE} = 3kT/2$$

Here Average KE is the average kinetic energy of a single atom or molecule; k is Boltzmann's constant; and T is the Kelvin temperature. otential Energy (abbreviated PE) is energy of position. A rock poised on top of a hill, even though it is not moving, has the potential to move if given a smll push. It therefore has potential energy. This potential energy is converted to kinetic energy as therock rolls down the hill. Allenergy is either kinetic or potential. The total enery of any mechanical system is the sum of its kinetic and potential energies:

$$E = KE + PE$$

Potential energy is an abstract concept; it is difficult to understand how something can possess energy simply by virtue of its position. A weight rests on a platform at height habove the ground. The weight is connected by a rope, across a pulley, to another platform that carries a load. If we remove the weight from the platform, it will fall, and in the process will lift the loaded platform. But in lifing the load a distance h, the weight has done work (a form of energy) on the load. To do work, the weight must use energy.

It possesses the necessary energy to do the work by virtue of its position at distance h above the earth. Thus the concept of potential energy presupposes a force of attraction or repulsion. The weight falls to earth because there is a gravitational force of attraction between the earth and the weight. To move the weight from the ground to the platform initially, we would have had to use an amount of energy sufficient to overcome this force of attraction.

The amount of energy used to lift the weight is stored in the earth-weight system as potential energy. We retrieve this energy by allowing the earth and weight to come together. Potential encrgy is a consequence of a force. Our example has been based on the gravitational force because this is a familiar situation.

There are 3 forces other than gravitation. One of these is the electromagnetic force. This is the force by which electrons are attracted and held by the nucleus in an atom. The electromagnetic force is responsible for all interactions between atoms and molecules. It is far stronger than the gravitational force, which may be neglected in discussing atomic and molecular interactions (the electromagnetic force of repulsion between two electrons is 10^{42} times stronger than the gravitational force of attraction between them!).

The electromagnetic force operating at the atomic level is responsible for:

- The attraction of electrons to the atomic nucleus;
- The bonding together of atoms in a molecule;
- The attraction between oppositely charged ions in an ionic compound;
- The attraction between molecules that leads to liquefaction and solidification;
- The flow of electrical current;
- Your inability to push your finger or hand through a wall.

The magnitude of the electrical force is given by Coulomb's Law, equation.

$$F = q_1 q_2 / W \times r^2$$

q_1 and q_2 are the values of the two interacting electrical charges, in Coulombs, and r is the distance between them, in m. W is a constant that adjusts the units. Its value is 1.113×10^{-10} $C^2N^{-1}m^{-2}$. The force described by Coulomb's Law is responsible for all attractive forces between atoms, molecules, and ions. The potential energy associated with the Coulomb force is called the Coulomb potential energy.

Because force is the negative derivative of potential energy with respect to distance ($F = -dPE/dr$), the Coulomb potential energy can be obtained by integration of Coulomb's Law.

$$PE = q_1 q_2 / W \times r$$

When one of the charges is positive and the other negative, the potential energy is negative at all finite values of r.

Coulomb Forces and the Potential Well

Atoms, molecules, and ions exert attractive forces on one another because the atomic nuclei in one species attract the electrons in the other. These are called molecular forces, and the potential energy that results from the action of these forces is called molecular potential energy. We make several general statements about molecular forces and potential energy, and the relationship between them. The statements are equally valid for a pair of particles or a large collection of particles (here we are using he word, "particle", generically to represent an atom, a molecule, or an ion).

- Molecular forces are electrical;
- Molecular forces are attractive except at very close distances, and described by Coulomb's Law;
- The magnitude (strength) of the force between 2 particles varies with their distance of separation. The force is zero for an infinite separation, increases as the particles get closer, becomes a maximum when the particles just touch, and becomes strongly repulsive when the centers of the particles are separated by less than one particle diameter, due to interpenetration and repulsion of lectron clouds. The force has two components: one operates only at very small distance and is repulsive; the other acts at all distances and is attractive. Thus,

  $$F = q_1q_2/W\times r^2 + B$$

  B is a positive, repulsive force that operates only at very small distance.
- Action of the molecular forces draws the particles together (tends to ause them to aggregate) and results in a decrease in the potential energy of he pair (or collection) of particles. The force becomes stronger and the potential energy less until the particles touch.

  At distances less than 1 particle diameter, the force is repulsive and PE increases again. For all types of molecular forces, potential energy varies with distance of molecular separation, r. The plot is shaped like a valley, or well. It is usually called a *poential energy well.* The depth of the well is the enrgy that must be adde to increase the distance of separation between particles from r_1 to r_3 or greater.
- There are several types of molecular forces, which we will discuss shortly. All of them are characterized by a potential well of similar shape. However, the depth of the well increases with the strength of the forces.

Region: The particles, shown as two small circles just above the horizontal axis, are separated by large distances. Taking 2×10^{-10} m as a typical particle radius, the distances of separation in region 1 are 10 or more times larger. At large separation distance, molecular forces are nearly zero. For convenience, we take the potential energy to be zero when the forces are zero, so the PE curve coincides with the horizontal axis at large r. This situation of very large separation between atoms, molecules, or ions is a state of *deaggregation*. The gas phase of a pure substance is an example of a deaggregated state.

Region: The particles have approached closely enough to exert noticeable forces on one another. Action of the forces lowers the potential enrgy. The closer the particles approach (*i.e.*, the more aggregated they become), the stronger the forces and the lower the PE. The particles just touch. At this point, attractive forces are maximized and potential energy is at a minimum. This is an aggregated state. The solid phase of a pure substance is an example of an aggregated state.

Region: PE increases very rapidly as r decreases. In this region, the somewhat deformable electron clouds of the particles are being pushed into one another.

Since electrons repel one another, two particles that have interpenetrated exert a repulsive force on one another, and tend to fly apart. Region shows that when particles are forced together so that their centers are separated by less than one molecular diameter, repulsive forces, hence PE, rise very rapidly. This is the force that keeps you from falling through the floor.

The attractive forces are maximized and the PE minimized when the atoms, molecules, or ions just touch, at the bottom of the well. The distance corresponding to the potential energy minimum is the distance of closest approach. We have called this the aggregated state. In an ionic compound, such as KBr, this corresponds to a situation in which positive ions are surrounded by negative ions, and vice versa.

In a covalent compound, such as H_2O, the bottom of the well represents the situation in which the hydrogen atoms are covalently bonded to the oxygen atom, the bonds are separated by an angle of 104°, and the distance between the H and O atoms corresponds to r at the bottom of the well. A collection of He atoms or methane molecules can maximize the contactbetween themselves by aggregating such that a particular molecule touches as many other molecules as possible. This organization is referred to as close packing, and characterizes the solid state of a pure substance. PE is minimized in the aggregated state. Thus shows how potential energy changes as a substance is converted from the deaggregated to the aggregated state or vice versa. Notice that potential energy increases when an aggregated state is converted to a deaggregated one:

$$\text{Aggregated} \rightarrow \text{Deaggregated DPE} > 0$$

This is an idea that has broad applicability; we will encounter it many times. The phases (solid, liquid, and gas) of a pure substance fit the aggregation/deaggregation model very well. In the solid phase, the fundamental units of a substance are packed together in an orderly arrangement that minimizes potential energy. This is the aggregated state, in which intermolecular attractive forces are optimized. In the gas phase, the fundamental units are widely separated and experience only very weak forces from one another.

This is the deaggregated state. We might ask where the liquid state is on the diagram. From macroscopic observation, we know that the molar volume of a substace, *e.g.* water, in the liquid phase is very little different from the molar volume of its solid phase. These two aggregated (or condensed) phases have molar volume roughly 1000 times smaller than that of the gas phase. The liquid phase has distance of molecular separtion and potential energy only smewhat greater than those of the solid, at the point labelled r_2.

The condensed (aggregated) phases of matter and the conversions between the gas, liquid, and solid phases. Prior to that, it is important that you have an understanding of the types of forces that operate within and between molecules, and their relative strengths. This will enable you to understand, for example, why some substances exist as high-melting solids under normal conditions, while others exist as low melting solids, liquids, or gases.

Types of Molecular Forces

The nature and structure of a substance determine the type and strength of molecular forces that operate within it. We divide such forces into two broad categories. Intramolecular forces operate within the molecules or fundamental units of a substance. These are the covalent bonds within a molecule of a molecular substance, the forces between ions in an ionic compound, and the forces between atoms in a metal. Intermolecular forces operate between, rather than within, the molecules of a covalent substance; the atoms of a monatomic element; or the ions of one substance and the molecules of another. As a rule, intramolecular forces are much stronger than intermolecular forces. Both types of force, however, give rise to potential energy and a potential well. A listing of the important forces, arranged within these categories, follows:

Intramolecular Forces

- Attraction between positive and negative ions in a crystal of an ionic compound.

- Covalent bonds linking atoms in a network structure, such as that of silicon dioxide (sand).
- Covalent bonds in molecular substances.
- Metallic bonds

Intermolecular Forces

- Attractions exerted by one molecule of a molecular substance on another, such as the force of attraction between water molecules in ice.
- Attractions between atoms of the noble gas elements, helium through radon.
- Attractions between molecules of one substance and molecules of another, as when two liquids are mixed, or a molecular solid such as sugar is dissolved in a liquid.
- Attraction between molecules of one substance and ions of another, as when an ionic compound dissolves in a liquid.

In succeeding sections, we will discuss each of these molecular forces, and their relative magnitudes. We will begin with intramolecular forces.

Intramolecular Forces

The Ionic Bond

When two elements of widely different electronegativities combine, the result is an ionic compound. An example of an ionic compound is potassium bromide, KBr, in which the electronegativity difference between bromine, in group 17, and potassium, in group 1, is 2.14. Ionic bond formation involves transfer of an electron from the potassium atom, with its single valence electron and core charge, Z_{core}, of 1 unit, to he bromine atom, with its seven valence electrons and large Z_{core} of 7 units. The electron transfer generates a full valence shell for each species. The ionic bond can be understood in terms of three energy erms. To remove an electron from the potassium atom to generate the potassium ion requires input of the first ionization energy of potassium.

The process is shown below:

$$K \rightarrow K^+ + e^- \quad I_1 = 420 \text{ kJ/mole}$$

I_1 for potassium is relatively small due to the minimum core charge found for group 1 elements. To add an electron to the bromine atom to form the bromide io releases an amount of energy equal to the first electron affinity of bromine.

$$Br + e^- \rightarrow Br^- \ -A_1 = -325 \text{ kJ/mole}$$

The first electron affinity for bromine is substantial, because the large core charge of the bromine atom exerts a reasonable force on the extra electron. Despite this, to accomplish the electron transfer from potassium to bromine requires that 420–325 = 95 kJ/mole of energy be supplied. In other words, the electron transfer is substantially unfavourable! Thus we find that, contrary to a commonly held view of ionic bond formation, potassium does NOT "like" to transfer an electron to bromine; the process is unfavourable, and is not responsibl for the strength of the ionic bond. What, then, is?

Generation of positive and negative ions results in a strong attraction between them, according to Coulomb's Law. A single positive and negative ion are shown coming together to form an ion pair in reaction below.

$$K^+ + Br^- \rightarrow K^+Br^- \text{ (ion pair)}$$

Formation of one mole of ion pairs releases 589 kJ of energy, much more than was required to generate the ions. We conclude that it is the Coulomb attraction between oppositely charged ions that drives the formation of the ionic bond; this happens despite the need to form ions, which is unfavourable. In fact, ionic compounds do not consist of ion pairs of the type shown in reaction above. Rather, they consist of extended, highly organized aggregates in which positive and negative ions alternate. Figureshows both the ion pair and the lattice of potassium bromide.

The lattice is more stable than a collection of ion pairs because it surrounds each ion with ions of the opposite charge. This maximizes the attractive forces between positive and negative ions.

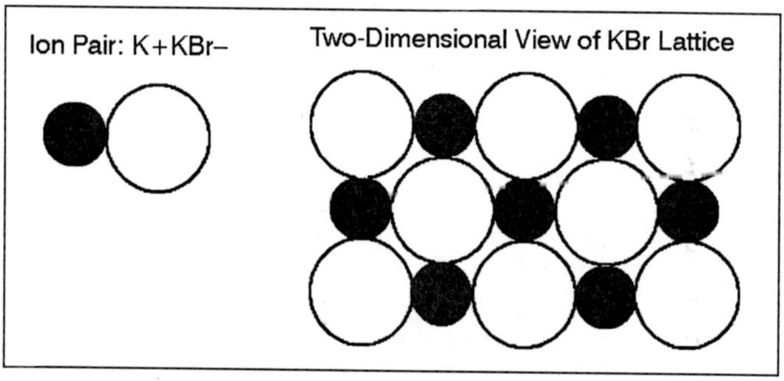

Figure: *Potassium Bromide, Ion Pair and Lattice*

In the ion pair, the positive charge of the potassium ion is offset on one side by the bromide ion, but is still exposed on all other sides. This exposed positive charge attracts bromide ions of other ion pairs, causing the ion pairs to condense to a lattice.

Formation of a lattice containing 1 mole each of potassium and bromide ions as in reaction releases 672 kJ of energy. This number, about 50 per cent larger than the amount released in formation of the equivalent amount of ion pairs, represents the depth of the potential well for potassium bromide.

$$K^+(g) + Br^-(g) \rightarrow KBr \text{ (lattice)}$$

The energy change that occurs when 1 mole of ionic substance in the solid lattice is formed from separated positive and negative ions in the gas phase is called the lattice energy, U, of the ionic compound.

The lattice energy is given by an expression having the form of:

$$U = KZ^+Z^-/r \text{ m}$$

Here K is a collection of physical constants, Z^+ is the cation charge, Z^- is the anion charge, and r is the distance of separation of the cation and anion in the lattice. Because the charge of the anion, Z^-, is negative, the lattice energy is a negative quantity, as expected.

This means simply that te solid lattice is more stable than the separated gas phase ions. Note that ths equation is essentially the expression for Coulomb potential energy. Without doing anycalculations at all, we can draw some important conclusions based on reaction.

- The potential well for an ionic compound of a 2+ cation with a 2- anion will be about 4 times deeper than that for a 1+/1- ionic cmpound, like KBr, due to the product of charges in the numerator of equation. Knowing the lattice energy for KBr allows us to estimate the lattice energy for MgO: about -2600 kJ/mole
- The potential well for an ionic compoun will be deeper when cation and anion are small than when they are large. Small cation and anionmeans small distance of separation, r, leading to a more negative U. The lattice energy for LiF is substantially more negative than that for KBr, despite the fact that both are 1+/1- compounds, ecause Li+ and F- are smaller ions. Lattice energies for a number of common ionic compounds are given in Table.

***Table:** Lattice Energies of Ionc Compounds*

Charge Type	*Substance*	*U, kJ/mole*
1 +/1–	LiF	–1009
	NaCl	–774
	KBr	–672
	CsI	–584
2 +/2–	MgO	–2585
	BaO	–1978

- *Example:* The lattice energy for NaCl is -774 kJ/mole. The radii of the sodium and chloride ions are 116 pm and 167 pm respectively. Estimate the lattice energy for CsI, which has ion radii of 181 and 206 pm.
- *Solution:* NaCl and CsI are both 1+/1- ionic compounds. They should differ in lattice energy only through the r term of 6-4-5. Because the distance of separation is lager in CsI, the lattice energy should be less negative. We assume that r can be approximated by adding the radii of the ions. Thus for NaCl, r = 283 pm; for CsI, r = 387 pm. The lattice energies should be inversely related to the distance of separation of the ions:

  U(CsI)/U(NaCl) = r(NaCl)/r(CsI)

  U(CsI)/(–774) = 283/387

  U(CsI) = –570 kJ/mole

This compares reasonably well with the value in Table.

Example. Arrange the following ionic compounds in order of increasing intramolecular forces: KI, LiCl, MgO, CaO Al_2O_3.

Solution. KI and LiCl are 1+/1- ionic compounds. MgO and CaO are 2+/2- ionic compounds. Finally, Al_2O_3 is a 3+/2- compound.

Based on ion charges and equation, we can arrange these in three categories:

- KI, LiCl
- MgO, CaO
- Al_2O_3

The 1+/1- compounds have the least negative lattice energies. That for Al_2O_3 should be most negative, because the ion charges are largest. Within each category, the compound with the largest ions has the least negative lattice energy. This reasoning leads to the following order:

$$U(KI) > U(LiCl) > U(CaO) > U(MgO) > U(Al_2O_3)$$

Here the greater than sign (>) is applied in the algebraic sense, in which numbers are considered to become smaller as they get more negative. In this convention, -50 is greater than–100. There are very few compounds that are legitimately considered to be ionic; that is, in which electron transferfrom metal to non-metal is complete. They are limited to compounds of the group 1 metals (Li, Na, K, Rb, Cs) with fluorine, chlorine, and oxygen. Compounds of the group 2 metals with oxygen, sulfur, and the halogens for typical ionic lattices, and are usually considered to be ionic compounds. However, there is considerable covalent character in the bonding.

Based on our discussion of the ionic bond in this section, it is understandable why ionic compounds are invariably solids at room temperature, with high melting and boiling points. The strong, non-directional Coulomb attractions acting between oppositely charged ions in the lattice can be overcome only at very high temperature.

The Covalent Bond

Two atoms of relatively large and similar electronegatvities (*i.e.*, non-metals) tend to form covalent compounds rather than ionic compounds. In covalent compounds, atoms share rather than transfer electrons to achieve full valence shells, because both atoms have similar substantial values of the core charge. Large Z_{core} tends to make electron transfer unfavourable. The shared electrons are paired, and are usually considered to be localized between the bonded atoms. In this section, we will discuss the chemist's view of the covalent bond. Two separated chlorine atoms, each with 7 valence electrons distributed in the 3s and 3p valence orbitals.

Each Cl atom has single odd electron in a 3p orbital. It is these electrons that will be pared and shared in the covalent bond. For the two atoms to bond by sharing electrons, it is necessary that they establish a mutual region in which the shared pair resides.

They accomplish this by merging, or overlapping, the atomic orbitals containing the electrons to be paired. Because these atomic orbitals have distinct spatial directions, the bonded atoms must maintain a definite orientatin relative to one another in order to preserve the overlap of orbitals. Thus covalent bonds are directional. The operation of Coulomb's Law in the covalent bond is perhaps less obvious than in the ionic bond. The two chlorine nuclei and the two electrons of the covalent bond, and indicates the various Coulomb forces that operate in this collection of centers.

There are several interactions: attrction of nucleus 1 for electron 1; attraction of nucleus 2 for electron 1; attraction of nucleus 1 for electron 2; attraction of nucleus 2 for electron 2; repulsion of nuclei 1 and 2; and repulsion of electrons 1 and 2. Note, however, that because the bonding electrns are required to be between the nuclei so that they can occupy the overlapping atomic orbitals, the nucleus-electron attractive forces are maximized, and the nuclear-nuclear repulsions are minimized, because the electrons shield the two nuclei from each other. Thus the strength of the covalent bond is attributable to the simultaneous attraction of both electrons to both nuclei.

The lengths and strengths of covalent bonds—the position and depth of the covalent bond potential well—vary with the bonded atoms. In general, thestrengths are substantial and comparable with those of ionic bonds. Thus to break all of the covalent bonds in a molecule of methane, CH_4, requires the input of 1662 kJ of energy. Methane lies in a deep potential well relative to the separated atoms. Generally speaking, the following generalizations hold for covalent bond strengths.

- Small atoms form relatively short bonds, and large atoms form relatively long bonds.
- Short bonds tend to be stronger than bonds formed by larger atoms because the bonding electrons are closer to the bonded nuclei. For example

Bond	*Length*	*Strength*
Cl–Cl	198 pm	240
Br–Br	228 pm	190
I–I	266 pm	149

- The word "tend" is emphasized here, because there are exceptions. Notably, single bonds of the small atoms, N, O, and F are very weak.
- Multiple bonds are stronger than single bonds. For example, the single, double, and triple bonds between carbon and oxygen have the following strengths:
  - C–O 358 kJ/mole
  - C = O 799 kJ/mole
  - C(3)O 1071 kJ/mole

This seems reasonable, because there are more electrons localized between the nuclei and attracted by both. Strengths and lengths of some typical covalent bonds are presented in Table. A more complete set of covalent bond strengths is found in Table below.

***Table:** Strengths and Lengths of Covalent Bonds*

Bond	*Bond Energy (Strength), kJ/mole*	*Bond Length, pm*
Single Bonds		
H – H	432	74.2
H – F	565	91.8
H – Cl	428	127.4
H – C	411	109
H – N	386	101
H – O	459	96
C – C	346	154
C – F	485	135
C – Cl	327	177
C – O	358	143
C – N	305	147
N – N	167	145
N – O	201	140
N – F	283	136
O – O	142	148
O – F	190	142
Multiple Bonds		
C = C	602	134
C = O	799	120
C = N	615	
N = N	418	125
N = O	607	121
O = O	494	120.7
S = S	425	188.7
C(3)C	835	120
C(3)O	1072	113
C(3)N	887	116
N(3)N	942	109.8
P(3)P	481	189
As(3)As	380	

- *Example*: Arrange the following single bonds in order of increasing length. Then arrange them in order of increasing strength.

H–O, O–Cl, Cl–Cl, H–Cl

- *Solution:* Longer bonds are formed by larger atoms; atoms become larger as we descend the periodic table. The sizes of the atoms involved here are ordered as follows:

Cl > O > H

The bond lengths are then expected to be in this order: H-O < H-Cl < O-Cl < Cl-Cl. The strengths of bonds tend to decrease with increasing length.

Thus:

Cl-Cl < O-Cl < H-Cl < H-O.

Although simplified, this section gives us an appreciation for the origin and strengths of covalent bonds. We now turn to a problem with our view of the covalent bond having to do with molecular stereochemistry. *The prediction and rationalization of molecular stereochemistry (shape)*: The VSEPR theory. Thus VSEPR predicts the tetrahedral geometry for methane, in exact agreement with experiment. The interpretation of the covalent bond in terms of overlap of atomic orbitals of the bonded atoms requires that we understand molecular shape in terms of the distribution in space of the atomic orbitals of the central atom, which are presumably overlapped with suitable terminal atom orbitals in covalent bonds.

We pose this question: are the observed shapes of molecules consistent with the distribution in space of the atomic orbitals of the bonded atoms? If we ask this question of methane, the answer must be NO. On one hand, the C-H bonds are known to make 109.5 ° angles with one another. On the other hand, the s and p valence orbitals of carbon have spatial orientations that are not consistent with this angle. The s orbital is spherical, hence non-directional; and the p orbitals make mutual angles of 90°. In fact, the bond angles in most molecules in which 3, 4, or 5 electron groups surround the central atom are inconsistent with this 90° orbital orientation. This presents a problem for our view of the covalent bond. This problem was recognized in the 1920s, when the modern quantum theory was applied to the hydrogen atom and the spatial shapes of the atomic orbitals were discovered. Tetrahedral stereochemistry at carbon in methane had been proposed in 1874 by. Vant Hoff on the basis of the number of isomers produced when hydrogen atoms were replaced by different atoms. At that time, though, people knew nothing of atomic orbitals. Thus the new quantum theory was faced almost immediately with a challenge: explain the shape of methane. This problem was tackled and resolved by Linus Pauling, who made during his life almost innumerable substantial contributions t science.

Pauling proposed that the carbon atom uses hybrid orbitals to form four identical bonds to hydrogen atoms, directed to the vertices of a tetrahedron. His theory of hybrid atomic orbitals is still widely used today to rationalize (not predict) molecular stereochemistry. Applied to methane, the theory goes as follows.

In its most stable electron configuration, the carbon atom. The s orbital is fully occupied with an electron pair, and two of the three p orbitals are singly occupied. Based on this arrangement, we might expect carbon to form two bonds using the two odd electrons. However, this would fail to give it the octet. Thus we imagine that carbon moves, or "promotes", one of the 2s electrons to the empty p orbital. This prepares it to form 4 bonds.

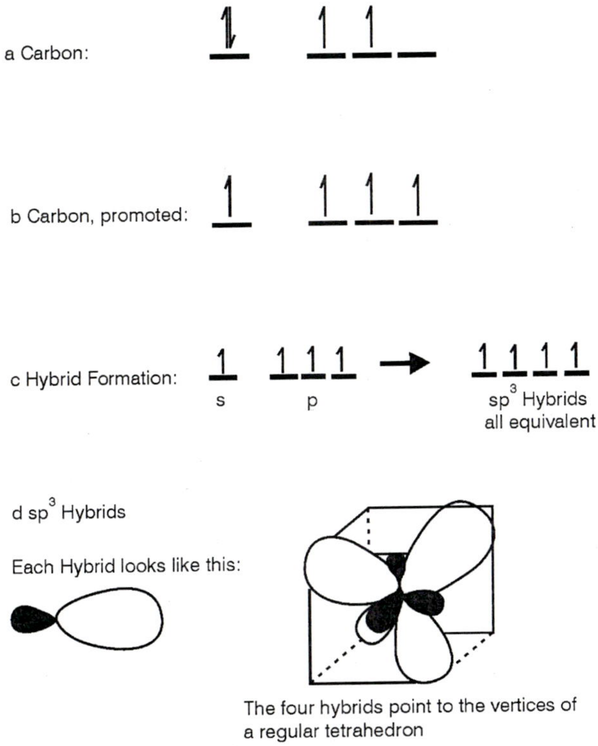

Figure: *The Electron Configuration of Carbon*

However, bond formation obviously does not occur using the pure atomic orbitals, because these can not produce a tetrahedral stereochemistry when overlapped with 1s orbitals of hydrogen atoms. Pauling proposed that the carbon atom creates from its 4 valence orbitals four new, equivalent orbitals called "hybrids." *The word hybrid means the same thing here as in agriculture*: A blend or combination of two strains of a species. He called these sp^3 hybrids to indicate that they are composed of a mixture of a single s and three p orbitals. Hybrid formation is shown schematically in figure.

Indeed, he showed mathematically that it is possible to combine the four s and p atomic orbitals to produce four new orbitals, and that these new orbitals are directed to the corners of a tetrahedron. The four hybrids are shown in Figure above. Pauling thus "squared" fact with quantum theory by proposing hybrid orbitals. Let's apply these ideas to a few more situations.

The phosgene ($COCl_2$) molecule is pictured in Figure. It has been experimentally determined that the molecule is trigonal planar (as expected from VSEPR), with a double bond between carbon and oxygen. How is the trigonal planar stereochemistry, involving 120° bond angles, to be rationalized with the 90° spatial distribution of carbon 2p orbitals?

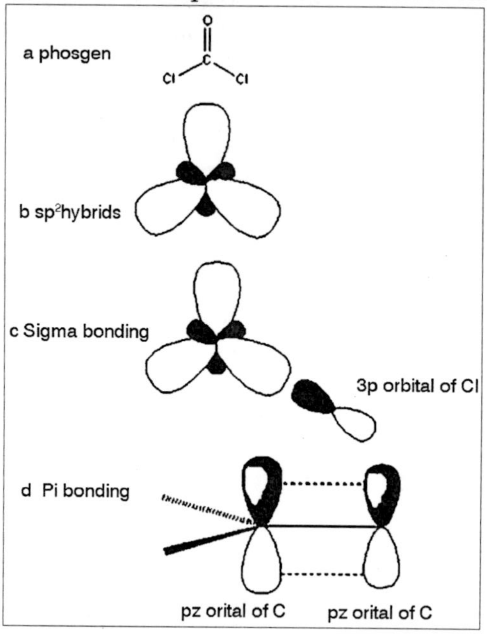

Figure: *Bonding in Phosgene*

Pauling showed that the 120° bond angle results if one assumes that C hybridizes the s and two of the three p orbitals to produce sp^2 hybrids.

Mathematically, the mixing of an s and two p's does indeed produce three equivalent new orbitals, directed to the corners of an equilateral triangle.

The plane in which the 3 hybrids lie depends upon which two p orbitals are used in the process. By convention, chemists normally choose to use px and py in the hybrids, which requires that the hybrids lie in the x-y plane of a Cartesian coordinate system. The pz orbital, not used in the hybrids, retains its usual shape and orientation along the z axis. Note, however, that it still contains an odd electron.

Once the hybrids are formed, each containing a single electron, carbon forms covalent bonds to the two hydrogen atoms and the oxygen atom. Each bond is formed by overlap of one carbon hybrid with one atomic orbital of the bonded atom (H or O). In forming these three bonds, orbitals are overlapped aong the line joining the bonded nuclei. Bonds involving overlap of this type are called *siga bonds*. At this point, bonding is not complete because carbon shares in only 7 electrons, as does oxygen. Each can complete the octet by forming one more bond, using the p orbitals oriented along the z axis.

Because these do not point at one another, but instead are parallel, overlap must occur in sideways fashion. Bonds formed by sideways overlap similar to this are called *pi bonds*. At this point, bonding is complete. We say that in phosgene, carbon uses sp^2 hybrids to sigma bond to H and O, and in addition forms a pi bond to O. The sp^2 hybrids are consistent with the trigonal planar sterochemistry. We are ready to make a few useful generalizations about hybrid orbital theory. Having done that, we will look at an example.

- The hybridization of an atom cannot be determined unless its stereochemistry is known. Hybrid orbitals cannot be used as a predictor of stereochemistry.

  CH_4 is tetrahedral at carbon; therefore, carbon uses sp^3 hybrids. NOT carbon uses sp^3 hybrids; therefore CH_4 is tetrahedral at C.
- The number of hybrids obtained is equal to the number of atomic orbitals used to make them. Thus, for example, there must be three sp^2 hybrids because three atomic orbitals (one s and two p) are used.
- An atom uses hybrid orbitals for its sigma bonds and lone pairs. Thus the number of hybrids required is equal to the sum of the number of lone pairs and sigma bonds:
- 4 sigma bonds + 0 lone pairs requires 4 hybrids, thus sp^3 (1 + 3 = 4)
- 3 sigma bonds + 1 lone pair requires 4 hybrids, thus sp^3 (1 + 3 = 4)
- 2 sigma bonds + 2 lone pairs requires 4 hybrids, thus sp^3 (1 + 3 = 4)
- 3 sigma bonds + 0 lone pairs requires 3 hybrids, thus sp^2 (1 + 2 = 3)
- 2 sigma bonds + 1 lone pair requires 3 hybrids, thus sp^2 (1 + 2 = 3)
- 2 sigma bonds + 0 lone pairs requires 2 hybrids, thus sp (1 + 1 = 2)

- Unhybridized p orbitals are used for pi bonds. The number of pi bonds equals the number of unhybridized p orbitals.

The flow of logic in determining hybrids used by the central atom is as follows. First, count the number of electron groups around the central atom; then apply VSEPR to determine the stereochemistry at that atom. The number of hybrids is the same as the number of electron groups counted for VSEPR.

A number of atomic orbitals equal to the required number of hybrids is used in forming the hybrids. The sum of the exponents in the hybrid designation indicates the number of hybrids of that type, and the number of atomic orbitals of each type used to make them.

Example: For each molecule, determine the Lewis structure; use VSEPR to determine molecular shape; and state the hybrid orbitals used by the central atom in bonding to the terminal atoms: $SiCl_4$; NH_3; BF_3; CO_2.

Solution: Lewis structures, then

Molecule	*#groups (hybrids)*	*Stereochemistry*	*Hybrids*
$SiCl_4$	4	Tetrahedral	sp^3
NH_3	4	Trigonal pyramidal	sp^3
BF_3	3	Trigonal planar	sp^2
CO_2	2	Linear	sp

Note that the number of pi bonds in BF_3 and CO_2 is the same as the number of unhybridized p orbitals on the central atom. It is relatively simple to extend the hybrid orbital concept to rationalize the stereochemistries of molecules in which the central atom has an expanded valence shell (more than 8 electrons). Essentially, this is done by incorporating in the hybrids as many valence d orbitals as required to give the necessary number of hybrid orbitals.

Example: Determine the hybrid orbitals used by the central atom in each molecule: PF_5, ClF_3, $XeOF_4$.

Solution:

Molecule	*# Electron Groups*	*Shape*
PF_5	5	Trigonal bipyramidal
ClF_3	5	T
$XeOF_4$	6	Square Pyramidal

Based on the numbers of groups, we deduce the hybrids. Five groups require 5 hybrids, which in turn require 5 atomic orbitals: one s, three p, and one d. Thus sp^3d. Six groups require 6 hybrids, which require the input of 6 atomic orbitals: one s, three p, and two d. Thus sp^3d^2. It can be mathematically demonstrated that 5 sp^3d hybrids point to the vertices of a trigonal bipyramid; and that 6 sp^3d^2 hybrids point to the vertices of an octahedron.

Before we leave the covalent bond and the hybrid orbital idea, something must be made clear. What is experimentally known is that carbon and hydrogen react to form methane, and that the methane molecule involves four carbon-hydrogen bonds that are perfectly tetrahedrally distributed in space. We introduced the ideas of electron promotion and orbital hybridization above as a way of rationalizing for ourselves, in terms of the output of quantum theory, what is experimentally observed. Thus promotion and hybridization are mental constructs. It is important that you realise that the carbon atom does not actually do these things. It is simply convenient for us to think of the carbon atom as doing these things.

There are numerous elements and compounds that consist of huge numbers of atoms covalently bonded together to give a large, 2- or 3-dimensional network. These substances are therefore similar to ionic compounds in that they do not consist of small, discrete molecules; however, they do not consist of discrete ions either. Examples of such substances are silicon dioxide (SiO_2), arsenic trioxide (As_2O_3), carbon in the diamond form, and phosphorus in the red form.

The large numbers of covalent bonds in these structures leads to potential well depths comparable to those found in ionic compounds. Most such materials exist as solids at ordinary temperature, and melt and boil only at very elevated temperature.

Ionic and Covalent Bonding: A Continuum

The bond between two identical non-metal atoms, as in Cl_2, is purely covalent; that is, each atom has an exactly equal share in the bonding electron pair. Pure covalent bonds exist in the elemental forms of many of the non-metals (for example, all of the diatomic halogens of group 7; O_2 and S_8 of group 6; N_2 and P_4 of group 5; C and Si of group 4; and others). Bonds between different non-metal atoms are covalent, in that they involve electron sharing, but polar, meaning that they have some ionic character. The bonds between chlorine and the other elements of period 3 are shown below. The transition from pure covalent to polar covalent to ionic as we move from right to left in the period is clear.

Na-Cl—Mg-Cl—Al-Cl—Si-Cl—P-Cl—S-Cl—Cl-Cl

ionic—ionic—ionic—polar cov—polar cov—polar cov—cov

Although the Na-Cl and Mg-Cl bonds are usually considered to be ionic, they are not purely ionic; electron transfer from the sodium or magnesium atom to the chlorine atom is not complete. There is no such thing as a pure ionic bond. The process of electron transfer from a metal atom to a non-metal atom generates a positive ion and a negative ion. The positive ion is electron deficient, and attracts an electron pair of the negative ion strongly. Thus the very act of electron transfer creates a disparity that must then be offset by partial electron sharing. If the anion is one of the very electronegative non-metals (flurine or oxygen), it can resist the pull of the cation effectively, and hold on to the electron pair.

Anions other than F^- and O^{2-}, however, cannot completely resist the attraction of the cation for their lone pairs. These atoms share a lone pair with the cation to form a bond that, although it is very polar, is nonetheless partially covalent. As a general rule, the higher the charges on the cation and anion, the more covalency in the bond. A highly charged cation (3^+ or 4^+) is small and compact, and exerts a powerful attractive force on electrons. A highly charged anion (2^- , 3^-) holds its extra electrons weakly and readily shares them to some extent with the cation. The following series of compounds illustrates nicely the evolution from largely ionic to covalent bonding as cation charge increases:

Compound	*Physical State, 298 K*	*Melting Point °C*	
NaCl	Solid	801	
$MgCl_2$	Solid	714	
$AlCl_3$	Solid	190	
$TiCl_4$	Iiquid	-25	

The decrease in melting point indicates increasing covalency in the bonding. Titanium tetrachloride, $TiCl_4$, is a liquid at room temperature, indicating that it is molecular, not ionic. The Metallic Bond. Elements of similar but relatively low electronegativities (*i.e.*, metals) form metallic bonds. It is these bonds that are operative in a typical metal and that are responsible for metallic properties: reflectivity, conductivity, malleability, and strength. Metallic bonds form between atoms that have fewer valence electrons than they have valence orbitals. We shall try to develop a picture of the metallic bond by considering sodium metal, which exists as a very low melting, highly reactive solid at room temperature. Figure shows the Lewis dot symbols for two sodium atoms.

The single odd electron is in the 3s valence atomic orbital of the sodium atom. At first glance, it might be expected that two sodium atoms, each with a single unpaired valence electron, would form a single covalent bond by overlapping the 3s valence orbitals and pairing the odd electrons in the resulting common region of space. There are problems with this bonding situation however.

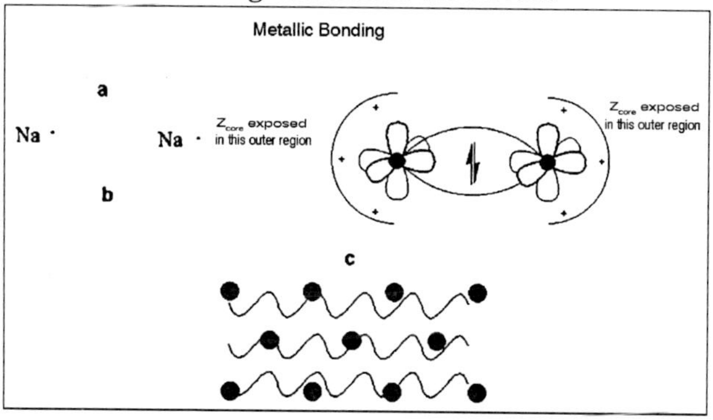

Figure: *Metallic Bonding*

First, bonding of this type would give each sodium atom only a duet of electrons, nowhere near the required octet. Each atom would have three empty 3p orbitals protruding from it into space. Second, pairing the 3s valence electrons localizes them in the region between the two nuclei.

They no longer have probability of being found in the peripheral regions opposite the bond. Because these are the only valence electrons, the regions marked with + symbols in the figure are exposed to the full core charge (+1) of the sodium atom. (The + signs in the figure do not represent actual positive charges; instead, they are intended to indicate that another atom located in this region would experience the core charge of the sodium atom near it). Here is the key point: another sodium atom in the region of the bonded pair would be drawn, via its valence electron, to the exposed core charge of the 2-atom cluster. It would contribute its valence electron to a bond with one or the other of the pair.

The sodium atoms are therefore driven to aggregate, using their relatively few valence electrons to bond together as many nuclei as possible via the Coulomb attraction. The aggregation is very regular in nature, resulting in an ordered arrangement of metal atoms called a *lattice*. As a result of aggregation, each sodium atom uses all of its valence orbitals in forming partial bonds with a number of neighbours.

Contrast this with the covalent bonding between two chlorine atoms. The covalent bond is again localized, but the remaining valence electrons are distributed evenly around each chlorine atom, effectively shielding the core charge from chlorine atoms in other molecules. Further, all of the s and p valence orbitals are full. In order to aggregate as sodium does, chlorine would have to use either the 3d orbitals or the orbitals from the n = 4 main shell. But the core charge experienced by an electron in either of these orbital sets is essentially zero (17-17 = 0). Consequently, chlorine is limited in its aggregation to the formation of covalently bonded diatomic molecules. As we will discuss shortly, there is only a very weak tendency for the diatomic molecules to aggregate via intermolecular forces. Interestingly, although hydrogen, like sodium, has only a single valence electron, it is not metallic. Instead, hydrogen atoms form covalently bonded diatomic molecules.

Why does hydrogen behave so differently from sodium and the other metals? Formation of a localized covalent bond between two hydrogen atoms results in exposure of the peripheral regions of the molecule to the core charge of the hydrogen atoms. In this case, because there are no core electrons, the H atom core charge is the nuclear charge of 1+. In the case of sodium and other metals, this situation leads to aggregation using the empty valence p orbitals.

However, hydrogen has no valence p orbitals. The single orbital in the n=1 shell is full; no further aggregation is possible. The formation of metallic bonds can occur only if empty valence orbitals are available. Experiments have shown that if hydrogen is subjected to extremely high pressure, which forces the diatomic molecules to aggregate, it exhibits metallic characteristics.

In sodium, as in most metals, there are too few electrons to form normal two-electron bonds between all pairs of bonded atoms. To achieve stability, the valence electrons are spread over many nuclei, simultaneously bonding them all together. This spreading of electrons is called *delocalization*, and in some ways is similar to the delocalization that occurs in resonance. In metals, however, delocalization is much more extensive, and it is best to think of the electrons as waves that extend over the entire dimensions of the metallic crystal. A schematic structure of metallic bonding.

The nuclei are represented as points, and the electrons as waves. This picture enables us to understand the properties of metals. That the electrons are delocalized waves is consistent with the conductivity of metals, and with their flexibility and strength. Putting stress on metals moves the nuclei, but the waves adjust, maintaining bonding. Because the electrons are waves, they are everywhere in the crystal at once. They are mobile, rationalizing the electrical and thermal conductivity of metals. Finally, the pooling of valence orbitals by aggregated sodium atoms leads to a large number of closely spaced orbitals in the metallic crystal. Thus the electrons absorb and reemit photons of all energies in the visible region of the spectrum, giving metals their highly reflective appearance.

There are so few bonding electrons in metallic sodium that only about 1/8 of an electron pair bond holds each pair of adjacent sodium atoms together. It is little surprise, then, that sodium is very soft (weak bonding), very reactive (deficiency of bonding electrons), with low density, and low melting and boiling points.

Example: In magnesium, how many metallic bonds form per Mg atom?

Solution: The electron configuration of the magnesium atom is $1s^2\ 2s^2\ 2p^6\ 3s^2$. We assume that metallic bonds are formed using the valence electrons only. Because each Mg atom contributes 2 valence electrons to the metallic bonding, and because 2 electrons constitute one bond in Lewis terms, there is one bond per Mg atom in the metallic lattice.

Intermolecular Forces

Having described the forces that operate within molecules (including ionic crystals and metals), we are ready to discuss the interactions that occur between species (atoms, molecules, and ions) that are not covalently bonded to each other. These are called *intermolecular forces*; they can originate in a number of ways.

Dipole-Dipole Forces

Intermolecular forces that operate between neutral molecules having molecular dipole moments are called dipole-dipole forces. The structure of a molecule may cause its centers of positive and negative charge to be displaced from one another, giving the molecule a dipole moment. The dipole moments of two neighbouring molecules tend to align with the + end of one dipole near the–end of the other, so that forces of attraction between them are maximized.

The maximum force of attraction between two dipoles, m_1 and m_2, separated by a distance r is given by equation.

$$F = \mu_1 \times \mu_2/\Omega \times r^4$$

Equation is Coulomb's Law, in disguised form. It is necessary to raise the distance of separation to the 4th power in the denominator to cancel the distance units that appear in both dipole moments in the numerator. Physically, the 4th power dependence means that the attraction between dipoles falls off much more rapidly with distance than would the attraction between isolated charges.

This is because the + and–ends of dipole 2 are almost equidistant from dipole 1 when the dipoles are separated by a distance that is much greater than the dipole length. The force of attraction between the + end of dipole 1 and the–end of dipole 2 is almost cancelled by the force of repulsion between the two + ends. Dipole-dipole forces operate between molecules of water, and between molecules of the substances.

Such forces are obviously much weaker than those operating in ionic or covalent network solids, and give rise to potential wells having depths in the approximate range 5-20 kJ/mole. Many molecular substances with dipolar molecules exist as liquids at ambient temperature, and have relatively low boiling points. In particular, many organic (carbo-containing) compounds are of this type.

Example: Which substances experience dipole-dipole intermolecular forces?

$$SiF_4, CHCl_3, CO_2, SO_2$$

Solution: We can draw the following conclusions about the molecules of these substances.

SiF_4	tetrahedral	Si-F bonds are polar, but no molecular dipole; bond dipoles cancel
CO_2	linear	C-O bonds are polar, but no molecular dipole; bond dipoles cancel
SO_2	bent	S-O bonds are polar and do not cancel. Sulfur lone pair dipole only partially offsets net bond dipole.
$CHCl_3$	tetrahedral	C-H and C-Cl bonds are polar and do not cancel

Hydrogen Bonding

Hydrogen bonding (H-bonding) is a special kind of dipole-dipole force that occurs when a hydrogen atom is bonded to one of the very electronegative atoms, F, O, or N.

The H-F, H-O, and H-N bonds are very polar, because the electronegative atom draws the bonding electron pair strongly to itself.

This leaves the hydrogen nucleus exposed. The resulting bond dipole is substantial. The positive end of this dipole, consisting essentially of an exposed hydrogen nucleus, has a high charge density because the proton is so small.

Consequently, it strongly attracts the negative end of a similar bond dipole in another molecule to give a structure.

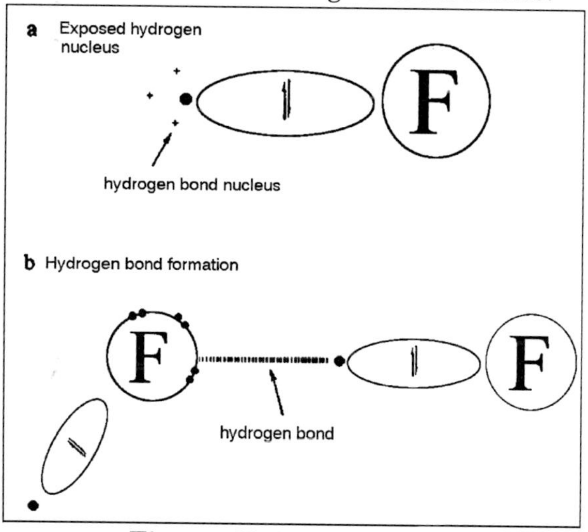

Figure: *Hydrogen Bonding*

The hydrogen atom is sandwiched between the two electronegative atoms in a linear arrangement. Hydrogen bond forces cause potential wells of depth in the range 5-50 kJ/mole.

Note that these are on average deeper than those resulting from ordinary dipole-dipole forces. The intermolecular forces between water molecules are of the hydrogen bonding type.

They are responsible for the abnormally high boiling point of water, which, with its small molecular weight, would otherwise be a gas at room temperature.

Compare the boiling point of water (100°C) with that of methane (-162°C), which has a similar molecular weight, but lacks hydrogen bonding and ordinary dipole-dipole forces. Hydrogen bonding is responsible for the formation of genes in the DNA molecule; for the helical structure of proteins; and for the incredible strength of Kevlar, a DuPont polymer used for canoe hulls and bullet-proof vests.

Example: In which substances do hydrogen bonding forces operate between molecules?

$$CHCl_3,\ CH_3CH_2OH,\ HNO_3,\ PH_3$$

Solution: Hydrogen is bonded to one of the very electronegative atoms in CH_3CH_2OH and HNO_3. Hydrogen bonding should occur in both of these substances.

Instantaneous Dipole-Induced Dipole (Dispersion) Forces

It may seem that there would be no forces of attraction between atoms of helium, because they are uncharged and have no permanent dipole moment. Yet He can be liquified at the very low temperature of 4K. That this is possible indicates that intermolecular forces do exist, though they are very weak. What is their origin? We portray the He atom as if the electron distribution about the nucleus is perfectly spherical. However, the electrons are constantly in motion and it is possible at any given instant for the electron cloud to be skewed slightly to one side of the nucleus.

This generates an instantaneous dipole moment in the He atom. During its transitory existence, this dipole induces a dipole in a neighbouring atom and the two instantaneous dipoles attract. Forces between transitory dipoles are called instantaneous dipole-induced dipole forces, or alternately, London dispersion forces after the scientist who first proposed them. We will refer to them as *dispersion forces.*

It is these forces that are responsible for the liquefaction and/or solidification of a number of substances whose molecules do not possess permanent dipole moments. For example, dry ice is solid CO_2. The CO_2 molecule is linear and, though it has non-zero bond dipoles, has no overall dipole moment.

Similarly, CCl_4, with perfectly tetrahedral molecules and no net dipole moment, is a liquid at room temperature and up to 78 °C. And iodine, which consist of I_2 molecules, is a solid at room temperature. Dispersion forces depend on two features of molecular structure. First, they increase in magnitude with the size and distortability (usually called the polarizability) of the electron clouds of the interacting particles. Size and polarizability increase as molecular weight increases.

It follows that dispersion forces increase as molar mass increases. For substances of large atomic or molecular mass, dispersion forces are strong enough that the substances are solid or liquid at room temperature.

We saw this above for CCl_4 and I_2. A striking example is provided by the diatomic halogens of group 17, which progress from F_2, a gas at room temperature and 1 atm pressure, to Cl_2, also a gas, to Br_2, a liquid, and finally to solid I_2. Only dispersion forces are operative in this series of substances. Second, dispersion forces depend upon molecular shape via the surface area over which two molecules can be in contact.

The larger the surface area of contact, the stronger the dispersion forces. Molecules that are roughly spherical in shape are able to contact each other only minimally. In contrast, molecules that are planar or linear in shape can maintain a large surface area of contact, with correspondingly larger dispersion forces. The classic example of the effect of molecular symmetry (shape) on the magnitude of dispersion forces is provided by the series of isomeric pentanes (C_5H_{12}).

Neopentane is highly symmetrical and nearly spherical in shape. Intermolecular forces are relatively small and the boiling point is correspondingly low. Pentane is linear, so two pentane molecules are able to contact each other along the entirety of their length. Intermolecular forces are relatively large, and the boiling point is correspondingly high. Isopentane is intermediate structurally, and the least symmetrical of the three isomers. Intermolecular forces are less than in pentane, but greater than in neopentane.

The boiling point is intermediate. Graphite, the most stable form of elemental carbon at ordinary temperature and pressure, has the structure. Each carbon atom is bonded to three others in trigonal planar fashion, using sp^2 hybrid orbitals.

The unhybridized p orbitals are used to form a network of pi bonds that extends over the entire array of bonded carbons, producing a large planar sheet of carbon atoms. These sheets are then stacked one atop the other to produce the three dimensional structure of graphite. Interestingly, there are no covalent bonds holding one sheet to the next.

The sheets attract one another via dispersion forces, which are very strong because of the large surface area of contact. It is important to realise that dispersion forces operate between all molecules, whether or not other forces also operate. Molecules of chloroform, $CHCl_3$, are atttracted by a combination of dipole-dipole and dispersion forces. The magnitude of dispersion forces varies with molar mass; however, they are generally weaker than dipole-dipole forces, giving potential wells in the range 0.1 to 5 kJ/mole.

Example: Arrange the following non-polar molecules in order of increasing melting point.

$$SiF_4,\ CS_2,\ CI_4,\ GeCl_4$$

Solution: None of these molecules is expected to have a dipole moment. Only dispersion forces will operate, and these increase with increasing molecular mass. The molar masses of these substances follow:

Substance	*MM*
SiF_4	104.077
CS_2	76.131
CI_4	519.631
$GeCl_4$	214.402

The intermolecular forces, and the melting points, should increase in the following order:

$$CS_2 < SiF_4 < GeCl_4 < CI_4$$

The experimentally determined melting points are -110.8, -90, -49.5, and 171 °C, respectively.

Example: Molar masses, dipole moments, and boiling points for several covalently-bonded substances are given below. Rationalize the observed boiling points in terms of intermolecular forces.

Substance	*MM*	μ	*Boiling Point, K*
He	4.003	0	4
H_2	2.016	0	20
N_2	28.014	0	77
CO	28.010	0.33	83
HCl	36.461	3.60	188
HBr	80.912	2.67	206
SO_2	64.058	5.42	263
HF	20.006	6.37	292
CCl_4	153.823	0	350
H_2O	18.015	6.17	373

Solution: The substances fall into two categories, those with dipole moments and those without. Only dispersion forces operate in substances without dipole moments.

Selecting from the table those substances with no dipole moment, shown below, we see that, with the exception of hydrogen, boiling point increases with increasing MM, as expected from our discussion of dispersion forces above. The dipole moment of CO is so small that it seems reasonable to include it in this list. Its boiling point is only slightly larger than that of N_2, due to weak dipole-dipole forces in addition to the dispersion forces.

Substance	*MM*	μ	*Boiling Point, K*
He	4.003	0	4
H_2	2.016	0	20
N_2	28.014	0	77
CCl_4	153.823	0	350

Substances having substantial dipole moments are listed below, now in order of increasing dipole moment.

Substance	*MM*	μ	*Boiling Point, K*
HBr	80.912	2.67	206
HCl	36.461	3.60	188
SO_2	64.058	5.42	263
H_2O	18.015	6.17	373
HF	20.006	6.37	292

We see only a rough correlation of boiling point with dipole moment: HBr and HCl have boiling points on the low end of the scale; SO_2 is intermediate; and H_2O and HF have high-end boiling points. However, the boiling points of HBr and HCl are inverted based on the size of the dipole moment. HCl, with the larger dipole moment, has the lowest boiling point in the group of five. Here is clear evidence that dispersion forces operate in addition to dipole-dipole forces.

Dispersion forces are substantially larger for HBr (MM 80.9) than for HCl (MM 36.5). This is sufficient to raise the boiling point for HBr above that for HCl. The remaining 3 substances do not yield to the same argument, however. The MM of SO_2 is more than three times larger than those of HF and H_2O, and its dipole moment is only somewhat less than that for water.

Yet the boiling points of HF and particularly of H_2O are strikingly higher than that for SO_2. Here we have evidence of the potency of hydrogen bonding intermolecular forces, which operate in both HF and H_2O. Despite its small MM, HF boils 30° higher than SO_2 because hydrogen bonding causes the potential well of the liquid to be relatively deep. The boiling point for water is about 80° higher than that for HF because each water molecule can engage in up to four hydrogen bonds (one for each H atom, and one for each lone pair on the O atom).

Mixed Forces in Solutions

All of the forces that can operate at the molecular level within a single pure substance are discussed above. It is possible to have what we might call hydrid forces that operate when pure substances are mixed to form solutions. When an ionic solid such as sodium chloride dissolves in water, ion-dipole forces are responsible. The maximum magnitude of ion-dipole forces is given by a form of Coulomb's Law that is intermediate between equations.

$$F = q \times \mu/\Omega \times r^3$$

Here q is the charge on the ion, m is the dipole moment of the solvent, and r is the distance between the ion and the centre of the solvent dipole.

Example: Consider the following three situations:

- A charge of 1+ separated by 2.0×10^{-10} m from a charge of 1- (this might be two ions in a crystal);
- A charge of 1+ separated by 2.0×10^{-10} m from a dipole of magnitude 1.6×10^{-29} C-m (a dipole of this magnitude would result from charges of 1+ and 1- separated by a distance of 1.0×10^{-10} m).
- Two dipoles, each of magnitude 1.6×10^{-29} C-m, separated by 2.0×10^{-10} m.

Compare the magnitudes of the attractive forces operating in these situations. This will provide an idea of the relative magnitudes of ion-ion, ion-dipole, and dipole-dipole forces. Then, compare the forces when the distance of separation of the interacting entities is increased by a factor of 4 to 8.0×10^{-10} m.

Solution: We apply equations:

- *Ion-ion*: $F = (1.6 \times 10^{-19}\ C)^2/W \times (2.0\times10^{-10})^2 = 6.4 \times 10^{-19}/W$ Newtons
- Ion-dipole, $F = (1.6 \times 10^{-19})(1.6 \times 10^{-19})/W \times (2.0 \times 10^{-10})^3 = 3.2 \times 10^{-19}/W$ Newtons
- Dipole-dipole $F = (1.6\times10^{-29})^2/W \times (2.0\times10^{-10})^4 = 1.6 \times 10^{-19}/W$ Newtons

The ion-ion force is stronger by a factor of 4 than the dipole-dipole force. The ion-dipole force is intermediate. When the separation distance is increased by a factor of 4, the magnitudes of all forces decrease because they depend inversely on a power of the distance. The ion-ion force is reduced by a factor of 4^2; the ion-dipole force by a factor of 4^3; and the dipole-dipole force by a factor of 4^4:

- *Ion-ion*: $F = 4\times10^{-20}/W$ NtIon-dipole $F = 0.5\times10^{-20}/W$ NtDipole-dipole $F = 0.063\times10^{-20}/W$ Nt

The forces involving dipoles clearly decrease much more rapidly with increasing distance of separation than do ion-ion forces. Whether or not two liquids are mutually soluble (miscible) depends on whether the mixed intermolecular forces can compete with the forces operating within each pure liquid alone. The intermolecular forces in pure water are primarily hydrogen bonding (dipole-dipole) forces, superimposed on weak dispersion forces.

Those in pure acetone are dipole-dipole forces superimposed on dispersion forces that are stonger than the dispersion forces in water. When acetone and water are mixed, they readily dissolve in one another in all proportions, apparently because the dipole-dipole forces between water and acetone molecules are of strengths comparable to those operating in the two separate liquids. On the other hand, only dispersion forces operate within pure carbon tetrachloride, because the CCl_4 molecule has a molecular dipole moment of zero. Because the molecules are spherical, giving small surface area of contact, these forces are fairly weak.

Consequently, carbon tetrachloride and water are not mutually soluble. They are said to be *immiscible*. When CCl_4 and water are mixed, the two liquids segregate into layers, with the denser liquid, chloroform, on the bottom. The liquid-liquid interface (boundary) is clearly visible. Finally, a non-polar solid like I_2 dissolves readily in a non-polar liquid like CCl_4, because the mixed dispersion forces are comparable in magnitude to the dispersion forces in pure CCl_4 and pure I_2. However, the solubility of I_2 in water is essentially zero, because mixed forces cannot compete with the strong dipole-dipole forces in water. Water molecules are unwilling to separate in order to make room for the non-polar I_2 molecules. In this way, we can extend our intermolecular force ideas to provide an understanding of the formation of mixtures.

Mixed-Force Concepts

- Substance A will dissolve in substance B if the forces operating between A and B molecules are of similar type (dipolar or dispersion) and strength to those operating between A molecules and between B molecules. This statement is often sloganized: "Like dissolves like."
- Substance A will not dissolve in substance B if the forces operating between A and B molecules are of different types (one dipolar, one dispersion) (and therefore usually different strengths) than those operating between A molecules and between B molecules. This can also be sloganized: "Like does not dissolve unlike."

Example: For each pair, predict whether substance A will dissolve in liquid B, and explain your reasoning.

Substance A	*Liquid B*
C_2H_5OH	CH_3OH
$Br_2(l)$	C_5H_{12}
$Br_2(l)$	H_2O
C_2H_5OH	H_2O
C_2H_5OH	$CHCl_3$
NaCl	$CHCl_3$
NaCl	CH_3CH_2OH
H_2O	CH_2Cl_2

Solutionl: C_2H_5OH (ethanol) and CH_3OH (methanol) are both liquids due to fairly strong H-bonding intermolecular forces. Ethanol is expected to dissolve in methanol. Br_2(l) and C_5H_{12} (pentane) are both non-polar liquids with dispersion forces operating between molecules. Thus they are "like" and we expect bromine to dissolve in pentane. Br_2 and H_2O are "unlike." Bromine will not dissolve readily in water.

Both liquids exert H-bonding forces at the molecular level. Ethanol is expected to dissolve in water.

This is not obvious. At first glance, we would call ethanol and chloroform "unlike" because H-bonding forces operate in ethanol, whereas dispersion forces operate in chloroform. However, in addition to its polar -OH group, ethanol has a non-polar hydrocarbon portion that is capable of exerting dispersion forces on molecules of chloroform. These two liquids are therefore miscible. Ionic NaCl will certainly not dissolve in non-polar chloroform.

Because ethanol is polar at the -OH end, we might expect NaCl to dissolve to some extent in ethanol, but not to the same extent as in water. This is indeed found. Polar water will not mix with non-polar dichloromethane.

Example: Describe what the system will look like during and following each process:

- 50 mL of distilled water is added to an Erlenmeyer flask.
- 0.10 g of I_2(s) is added to the flask, and the flask is stoppered and vigourously shaken.
- 50 mL of CCl_4 is added to the Erlenmeyer flask, and the flask is stoppered and shaken.

Solution:

- The clear, colourless liquid will occupy the bottom of the flask.
- I_2, and intensely violet solid, is not expected to be soluble in water because of incompatible intermolecular forces. Because I_2 is more dense than water, it will sink to the bottom of the flask. The water will remain clear and colourless.
- CCl_4 is a non-polar liquid in which the intermolecular forces are of the dispersion type. It is expected to be immiscible with water. Because it is more dense than water, it will sink to the bottom of the flask, and the water will float on it with a distinct liquid-liquid boundary. CCl_4 and I_2 have compatible intermolecular forces, so I_2 is expected to dissolve readily in

carbon tetrachloride to give a purple solution. The flask will thus contain a layer of purple liquid under a layer of colourless liquid.

Distribution of a solute Between Two Immiscible Liquids in Contact

When two immiscible liquids, such as water and chloroform ($CHCl_3$), are placed in contact, they will mutually exclude each other, forming two liquid layers. The less dense liquid will float on the denser one, and a boundary surface will be clearly visible. It is interesting to think about what will happen if a third substance, solid or liquid, is added to the system consisting of two immiscible liquids, A and B.

If the third substance, C, is soluble in one of the liquids (say A) and insoluble in the other (B), it will dissolve exclusively in the A layer, as in example 6-14. More interesting is the situation in which the C is soluble in both liquids to at least some extent. In this case, we find that the substance C *partitions*, or *distributes*, itself between the two liquid layers. We further find that the concentration of the third substance is higher in the liquid with which its intermolecular forces are most compatible. The ratio of the concentration of C in liquid A to that in liquid B is found to attain a constant value K_d, defined as in equation:

$$K_d = [C]_A/[C]_B$$

This constant ratio is established no matter how the 3-component mixture is prepared.

Three such ways follow:

1. A and B are first brought into contact. C is then added, and the container is shaken.
2. C is first dissolved in A. A (containing C) and B are then brought into contact, and the container is shaken.
3. C is first dissolved in B. A and B (containing C) are then brought into contact, and the container is shaken.

In addition, the ratio is reestablished if the mixture is altered in any of the following ways:

- More A is added.
- More B is added.
- More C is added.

We can understand this behaviour in terms of a simple layers of A and B in contact at a boundary, and molecules of C initially dissolved in liquid A. The boundary between the two liquids provides a "doorway" from one to the other through which C molecules, which have reasonable interactions with both A and B molecules, can pass.

Thus molecules of C will tend to move across the boundary from liquid A, where they are initially dissolved, into liquid B, in which there are initially no molecules of C. In other words, they begin to distribute themselves between the two liquids. After some time, enough C molecules will have passed into liquid B that the rates of passage of C from A to B and back will be the same. Once this situation is attained, there will be no further change in the concentration of C in the two layers. In other words, the constant ratio in equation will be attained. From the magnitude of K_d, we may easily deduce which liquid C prefers. If $K_d < 1$, C prefers B; if > 1, C prefers A.

Distribution of a substance between two liquid phases is an example of dynamic equilibrium, a situation in which two opposing processes (here, movement of C both ways across the boundary) occur at equal rates.

A Nod to Other Classification Systems

Classification of the various types of forces as "inter-" or "intramolecular" is not absolute.

Thus you may see the forces classified as follows:

- Covalent Bonds (in molecular and network covalent substances)
- Ionic Bonds
- Metallic Bonds
- Interunit forces
  - Ion-dipole
  - Dipole-dipole
  - Hydrogen bonds
  - Dispersion forces

In this classification scheme, ionic and metallic bonds are categorized separately rather than as intermolecular forces, because there are no discreet molecules in ionic compounds or metals. Another classification system has recently become popular and is experiencing increasing use. This goes as follows.

- Covalent Interactions (includes covalent bonds in molecular and network covalent substances)
- Non-covalent Interactions
  - Ionic bonds
  - Metallic bonds
  - Metal-ligand interactions

- ion-dipole forces
- dipole-dipole forces
- hydrogen bonds
- dispersion forces

Here there are only 2 major categories: covalent, and not covalent! It is important to realise that the distinction between covalent bonding and some of the non-covalent interactions is blurred.

For example, many people believe that metal-ligand interactions and even hydrogen bonding are weak covalent interactions.

The bottom line is that it is not particularly important how the classification is done, as long as the various types of interactions are understood. Pick your own favourite classification system, and use it!

The Interplay Between Molecular Kinetic and Potential Energies

The concept of temperature is of profound significancein science. We take it for granted in our daily lives, without thinking much about what it is. Temperature is the manifestation of molecular kinetic energy. The steam rising from a pan of boiling water feels hot because the molecules of water vapour are moving rapidly. Their collisions with the skin transfer kinetic energy, which we interpret as heat. The relationship between temperature and the average translational kinetic energy of an atomic, ionic, or molecular unit is very simple.

$$\text{Average } KE_{molecule} = 3/2\ kT$$

T is the absolute (Kelvin) temperature, and k is a fundamental constant of nature called Boltzmann's constant. Its value is 1.381×10^{-23} J/K. By $KE_{molecule}$, we signify the kinetic energy of any microscopic unit of matter.

This can be anything from an electron to a DNA molecule. Keep in mind in using equation that the kinetic energy referred to is the translational kinetic energy—the energy of motion of the particle as a whole through space. Translational kinetic energy is often called thermal energy.

Example: Calculate the thermal energy of one mole of nitrogen gas at room temperature, 293 K.

Solution: Calculate the kinetic energy of an average molecule; then multiply by Avogadro's Number.

Average $KE_{molecule}$

$= 3kT/2$

$= 3(1.381 \times 10^{-23}\ \text{J/K})(293\ \text{K})/2$

$= 6.07 \times 10^{-21}$ J/molecule

KE_{mole}

$= N_o KE_{molecule}$

$= 3N_o kT/2$

$= 6.07 \times 10^{-21}$ J/molecule$\times 6.023 \times 10^{23}$ molecules/mole

$= 3.66 \times 10^3$ J/mole

The product of Boltzmann's constant and N_o is a constant, R, which occurs in the ideal gas law. Its value is 8.314 J/K-mole. Expressed in terms of R, the kinetic energy per mole of gas is 3RT/2. We have seen that molecular forces promote aggregation by lowering molecular potential energy. Kinetic energy counteracts the effects of intermolecular forces. These ideas lead to a very useful generalization:

Potential energy favours aggregation; kinetic energy opposes aggregation. The stronger the molecular forces, the more rapidly atoms, molecules, or ions must move in order to overcome them. Even the strongest forces, however, may be overcome by the vigour of molecular motion if the temperature is high enough. At a temperature that gives a kinetic energy comparable in magnitude to the depth of the potential well, molecules begin to escape the potential well, and the substance undergoes a phase change (that is, sublimes, melts, or boils).

Table: *Melting Temperature and Potential Well Depth*

Substance	*Potential Well Depth, J/mole*	*Melting T, K*
He	0.105	3.5
H_2	1.04	14.0
Xe	14.7	161
N_2	6.30	63.2
Cl_2	26.8	172.1
I_2	56.3	387
H_2O	46.7	273
CO_2	33.6	217
NaCl	771	1074

A phase change may therefore by viewed as an exchange of KE for PE. If the potential well is shallow, as in liquid helium (where the only forces are the very weak dispersion type) the required kinetic energy is achieved at very low temperature (4 K). If the potential well is deep, as in the ionic compound MgO (where the forces are the strong ion-ion type), very high temperature is required for particles to escape (3125 K). Table shows the parallel between the potential well depth and the temperature of melting for a number of substances.

Try plotting melting point against potential well depth to see how good the correlation is.

Non-Ideality of Gases

The postulates of KMT define an ideal gas. An ideal gas obeys the ideal gas law under all conditions of T and P. There is in fact no real gas that satisfies this definition. The ideal gas is a concept that we can think about but which does not actually exist.

Consider a sample of N_2 gas at 25 °C and 1 atm pressure. Imagine measuring the volume of the gas periodically as we cool it at constant pressure. If N_2 were ideal, it would follow Charles' Law exactly even to absolute zero. A plot of V versus T for ideal N_2 would follow curve A in Figure.

The actual V-T plot for N_2 follows curve A to a temperature of about 100 K. At this point the plot starts to curve down (curve B). At 77 K the volume sharply decreases while the temperature remains constant, and liquid appears in thecontainer. When all gas has converted to liquid, further decrease in T has little effect on the volume occupied by the liquid until T = 63 K. At this point, N_2 solidifies with a slight decrease in volume. Further cooling causes negligible change in V.

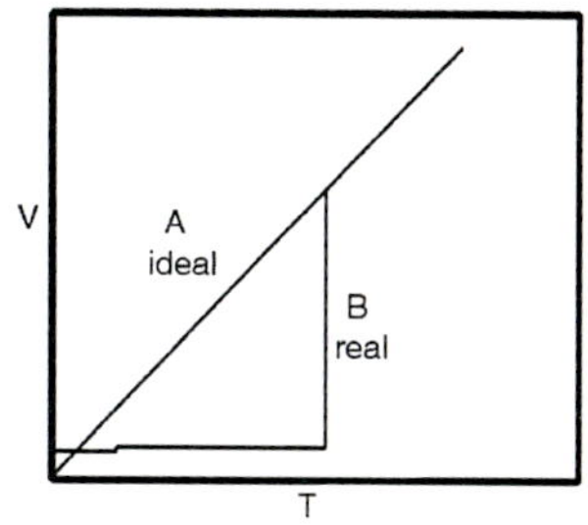

Because all real substances liquify at some T, no real substance can obey the ideal gas law under all P,T conditions. Deviation of real gases from ideal behaviour is shown conveniently in terms of the compressibility factor, Z

$$: Z = PV/nRT$$

Z is plotted against P in figure. For an ideal gas, Z = 1 at all pressures (the horizontal dashed line). Two deviations from ideality are commonly observed. In one type, Z always exceeds 1. H_2 is an example. In the second type, $Z < 1$ at moderate pressure but becomes > 1 at high pressure. N_2 is an example. Most gases show behaviour similar to that of N_2. Real gases deviate from ideality because they violate two postulates of the KMT.

First, real molecules exert attractive forces on one another. The forces cause the molecules to temporarily stick together in pairs when they collide, reducing the effective number of free particles in the gas. This is shown in Figure.

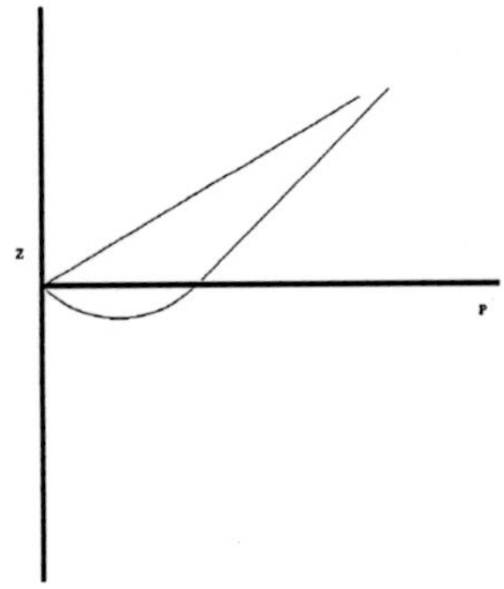

Figure: *The Compressibility Factor, Z*

Although there are 7 gas molecules in the container, 4 of them are temporarily stuck together in pairs, reducing the effective number of independent gas particles to 5. Sticking together reduces the pressure below the ideal value because there are fewer collisions with the container walls per unit time.

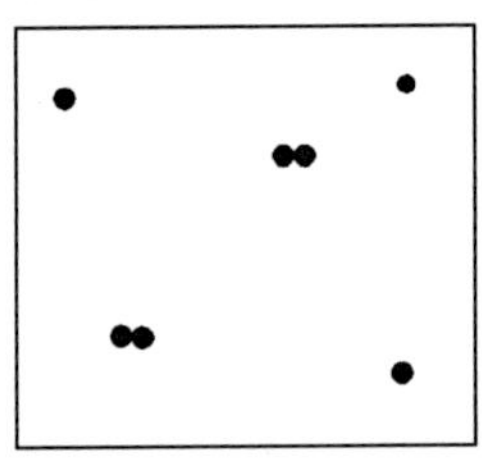

Figure: *Non-Ideality of Gases*

$$\text{Thus, } P_{act} = P_{ideal} - DP$$

Here P_{act} is the actual pressure of the gas, P_{ideal} is the pressure it would exert in the absence of intermolecular forces, and D P is the pressure reduction. If there are n moles of gas in a container of volume V, the pressure reduction, D P, is proportional to $(n/V)^2$.

The following argument demonstrates this. First, we assume that D P is proportional to the number of pairs of stuck-together molecules:D P proportional to (# pairs). How many pairs are possible for N molecules? If N = 2, there is only 1 pair. If N = 3, 3 pairs are possible. For N molecules, there are N(N-1)/2 possible pairs, where division by 2 avoids double counting. If N is very large, as in a macroscopic gas sample, N-1 = N and the number of pairs is $N^2/2$. Dividing by volume to convert to concentration and by N_0 to convert molecules to moles gives,

D P proportional to (# pairs) proportional to,

$$(n/V)^2 = a(n/V)^2$$

where "a" is a proportionality constant. Substitution in equation gives,

$$P_{ideal} = P_{act} + a(n/V)^2$$

Second, real molecules are not points, but instead occupy finite volume. Thus a small fraction of available container space is occupied by the molecules themselves, and is not available to accommodate motion of the molecules. The total container volume is the sum of the volume of the molecules and the free volume available for motion:

$$V_{container} = V_{available} + V_{molecules}$$

The volume of the molecules is constant for a given gas, and depends on molecular size. As $V_{container}$ is made smaller, $V_{available}$ becomes a smaller fraction of $V_{container}$. Since the molecules have smaller free volume than they would if ideal (when $V_{molecules} = 0$), they exert a larger than ideal pressure. Equating $V_{container}$ with V_{actual} and $V_{available}$ with V_{ideal}, we obtain,

$$V_{act} = V_{ideal} + V_{molecules}$$

The volume of the molecules is proportional to the number of molecules, which is in turn proportional to the moles of gas: $V_{molecules}$ proportional to n = bn where b is a proportionality constant roughly equal to the volume per mole of substance. (The volume of the molecules can be determined by crowding the molecules together—*i.e.*, by liquifying the substance. The molar volume of the liquid phase gives an estimate of "b". The molar volume of liquid is the product of the MW and the reciprocal of density: b(vol/mole) = MM(mass/mole)×1/r (vol/mass) = MM/r (vol/mole) Using this approach, b should be 18/1.0 = 18 mL/mole for water.) Substituting equation rearranging for V_{ideal} gives

$$V_{ideal} = V_{act} - bn$$

Finally we substitute equations into the ideal gas equation, $P_{ideal}V_{ideal} = nRT$ to obtain equation:

$$(P + an^2/V^2)(V - nb) = nRT$$

This equation was first proposed by van der Waals in 1873 to explain deviations of gases from ideal behaviour. Its success earned him the Nobel Prize for Physics in 1910. In real gases the effects of non-zero molecular volume and intermolecular forces oppose. The former causes pressure to be higher than ideal, and the latter causes it to be lower than ideal. For N_2 and many other gases having $Z < 1$ at moderate pressure and $Z > 1$ at high pressure, first one effect, then the other, takes precedence. At low pressure (10-100 atm), the intermolecular forces dominate because the molecules are not yet compressed enough for the volume effect to be felt. This makes $P_{act} < P_{ideal}$, and Z < 1. As P becomes very large (> 100 atm), non-zero molecular volume becomes important, making $P_{act} > P_{ideal}$ and Z > 1. For hydrogen, intermolecular forces are small and the volume effect dominates at all pressures.

At low pressure (approximately 1 atm) and temperature much higher than the boiling point of the substance, both non-ideal effects are small and real gases approach ideal behaviour. Values of the van der Waals constants a and b for several gases are in Table.

Table: *Van der Waals Constants for Gases*

Gas	*a, L^2-atm/mole*	*b, L/mole*	
Ar	1.35	0.0322	
Cl_2	6.49	0.0562	
CO_2	3.59	0.0427	
H_2	0.244	0.0266	
He	0.034	0.0237	
N_2	1.39	0.0391	
O_2	1.36	0.0318	

Example: Discuss the Van der Waals constants in Table in terms of molecular structure.

Solution: The constant, a, is related to the magnitude of intermolecular forces; b is related to molecular volume. As a first guess, we might expect both of these to get larger with increasing molar mass of the gaseous substance.

Let us predict that the constants will parallel molar mass. The substances and their constants are arranged in order of increasing MM below.

In a rough sense, our prediction is borne out. Both a and b trend upward as molar mass increases. However, we are wrong in the details. The value of a for He is a clear exception; and the constants for nitrogen, oxygen, and argon do not parallel the molar masses. Having predicted, somewhat but not entirely successfully, we now must rationalize (*i.e.*, try to think of explanations for deviations from the trend).

First, we examine the hydrogen, helium pair. The difference between these is that hydrogen is molecular, and helium is atomic. Hydrogen molecules are highly symmetrical, but elongated from a perfect sphere.

Gas	MM	a, L^2-atm/mole	b, L/mole
H_2	2.016	0.244	0.0266
He	4.003	0.034	0.0237
N_2	28.014	1.39	0.0391
O_2	31.998	1.36	0.0318
Ar	39.948	1.35	0.0322
CO_2	44.009	3.59	0.0427
Cl_2	70.906	6.49	0.0562

The hydrogen molecules are more susceptible to dispersion forces than are the spherical, compact (high core charge) helium atoms. Consequently, forces (and the constant a) are larger for H_2, despite its lower molar mass.

Incidentally, because there are 2 atoms per particle in hydrogen, the volume constant, b, is slightly larger for hydrogen. Next, we consider nitrogen, oxygen, and argon, which have roughly similar molar masses. Again, Ar is atomic and spherical, whereas nitrogen and oxygen are molecular, with elongated electron distributions. Despite their greater mass, the argon atoms are less susceptible to dispersion forces because they are spherically shaped. For the same reason, the volume constant for Ar is comparable to those for nitrogen and oxygen, despite the greater number of electrons. Atomic size decreases left to right in a period; oxygen atoms are actually smaller than nitrogen atoms, despite their larger atomic weight. Thus the volume constant, b, is smaller for oxygen than for nitrogen. Oxygen is more electronegative than nitrogen, making it somewhat less distortable. Intermolecular forces, measured by a, are somewhat less in O_2.

Electromagnetic Theory

James Clerk Maxwell, who had set himself the task of mathematically working out Michael Faraday's views, and who, both by doing so and by introducing many new ideas of his own, became the founder of the modern science of electricity,' recognized that, at every point of an electromagnetic field, the state of things can be defined by two vector quantities, the "electric force " E and the " magnetic force " H, the former of which is the force acting on unit of electricity and the latter that which acts on a magnetic pole of unit strength.

In a non-conductor (dielectric) the force E produces a state that may be described as a displacement of electricity from its position of equilibrium. This state is represented by a vector D ("dielectric displacement") whose magnitude is measured by the quantity of electricity reckoned per unit area which has traversed an element of surface perpendicular to D itself. Similarly, there is a vector quantity B (the "magnetic induction") intimately connected with the magnetic force H.

Changes of the dielectric displacement constitute an electric current measured by the rate of change of D, and represented in vector notation by C==D Periodic changes of D and B may be called "electric " and "magnetic vibrations." Properly choosing the units, the axes of coordinates (in the first proposition also the positive direction of *s* and n), and denoting components of vectors by suitable indices, we can express in the following way the fundamental propositions of the theory.

- Let *s* be a closed line, a a surface bounded by it, n the normal to a. Then, for all bodies, *H 8 ds =f C n dcr,* J *Esds =-d* f *Bnda,* where the constant *c* means the ratio between the electro-magnet and the electrostatic unit of electricity. From these equations we can deduce: For a surface of separation, the continuity of the tangential components of E and H. The solenoidal distribution of C and B, and in a dielectric that of D. A solenoidal distribution of a vector is one corresponding to that of the velocity in an incompressible fluid. It involves the continuity, at a surface, of the normal component of the vector.
- The relation between the electric force and the dielectric displacement is expressed by Dx = eiEx, Dy = e2Ey, D z = E3Ez, the constants (dielectric constants) depending on the properties of the body considered. In an isotropic medium they have a common value *e,* which is equal to unity for the free aether, so that for this medium D = E.
- There is a relation similar to between the magnetic force and the magnetic induction. For the aether, however, and for all ponderable bodies with which this article is concerned, we may write B =H.

It follows from these principles that, in an isotropic dialectric, transverse electric vibrations can be propagated with a velocity *v* = *c/-J E.* Indeed, all conditions are satisfied if we put Dx =o, Dy =*a* cos *n(t-xtr'+l), 'D z =o, H z =o, H y =0, Hz=* avc 1 *cos* n(t–xa+*l*) For the free aether the velocity has the value c.

Now it had been found that the ratio *c* between the two units of electricity agrees within the limits of experimental errors with the numerical value of the velocity of light in aether. (The mean result of the most exact determinations 2 of c is 3,001 to 10 cm./sec., the largest deviations being about 0.008 10 10 :. and Cornu 3 gives 3,001 to 10 0,003 to 10 as the most probable value of the velocity of light.) By this Maxwell was led to suppose that light consists of transverse electromagnetic disturbances.

On this assumption, the equations represent a beam of plane polarized light. They show that, in such a beam, there are at the same time electric and magnetic vibrations, both transverse, and at right angles to each other. It must be added that the electromagnetic field is the seat of two kinds of energy distinguished by the names of electric and magnetic energy, and that, according to a beautiful theorem due to J. H. Poynting, 4 the energy may be conceived to flow in a direction perpendicular both to the electric and to the magnetic force.

The amounts per unit of volume of the electric and the magnetic energy are given by the expressions,

2 (+ Ey Dy + Ez Dz),

a (HxBx +Hy B y+HzBz) = whose mean values for a full period are equal in every beam of light. The index of refraction of a body is given by result that has been verified by Ludwig Boltzmann's measurements' of the dielectric constants of gases. Thus Maxwell's theory can assign the true cause of the different optical properties of various transparent bodies.

It also leads to the reflection formulae, provided the electric vibrations of polarized light be supposed to be perpendicular to the plane of polarization, which implies that the magnetic vibrations are parallel to that plane. Following the same assumption Maxwell deduced the laws of double refraction, which he ascribes to the unequality of E l, 21 His results agree with those of Fresnel and the theory has been confirmed by Boltzmann,' who measured the three coefficients in the case of crystallized sulphur, and compared them with the principal indices of refraction. Subsequently the problem of crystalline reflection has been completely solved and it has been shown that, in a crystal, Poynting's flow of energy has the direction of the rays as determined by Huygens's construction.

Two further verifications must here be mentioned. In the first place, though we shall speak almost exclusively of the propagation of light in transparent dielectrics, a few words may be said about the optical properties of conductors. The simplest assumption concerning the electric current C in a metallic body is expressed by the equation C = QE, where a is the coefficient of conductivity.

Combining this with his other formulae. Maxwell found that there must be an absorption of light, a result that can be readily understood since the motion of electricity in a conductor gives rise to a development of heat. But, though Maxwell accounted in this way for the fundamental fact that metals are opaque bodies, there remained a wide divergence between the values of the coefficient of absorption as directly measured and as calculated from the electrical conductivity; but in 1903 it was shown by E. Hagen and H. Rubens that the agreement is very satisfactory in the case of the extreme infra-red rays.

In the second place, the electromagnetic theory requires that a surface struck by a beam of light shall experience a certain pressure. If the beam falls normally on a plane disk, the pressure is normal too; its total amount is given by c-1, if i 1, i2 and i 3 are the quantities of energy that are carried forward per unit of time by the incident, the reflected, and the transmitted light.

This result has been quantitatively verified by E. F. Nicholls and G. F. Hull.' Maxwell's predictions have been splendidly confirmed by the experiments of Heinrich Hertz' and others on electromagnetic waves; by diminishing the length of these to the utmost, some physicists have been able to reproduce with them all phenomena of reflection, refraction (single and double), interference, and polarization.

The wave-lengths observed in the aether now has Let the indices *p* and n relate to the two principal cases in which the incident (and, consequently, the reflected) light is polarized in the plane of incidence, or normally to it, and let positive directions *h* and *h* be chosen for the disturbance in the incident and for that in the reflected beam, in such a manner that, by a common rotation, h and the incident ray prolonged may be made to coincide with h and the reflected ray.

Mechanical Models of the Electromagnetic Medium

From the results already enumerated, a clear idea can be formed of the difficulties which were encountered in the older form of the wave-theory. Whereas, in Maxwell's theory, longitudinal vibrations are excluded *ab initio* by the solenoidal distribution of the electric current, the elastic-solid theory had to take them into account, unless, as was often done, one made them disappear by supposing them to have a very great velocity of propagation, so that the aether was considered to be practically incompressible. Even on this assumption, however, much in Fresnel's theory remained questionable. Thus George Green,' who was the first to apply the theory of elasticity in an unobjectionable manner, arrived on Fresnel's assumption at a formula for the reflection coefficient *A n* sensibly.

In the theory of double refraction the difficulties are no less serious. As a general rule there are in an anisotropic elastic solid three possible directions of vibration, at right angles to each other, for a given direction of the waves, but none of these lies in the wave-front. In order to make two of them do so and to find Fresnel's form for the wave-surface, new hypotheses are required. On Fresnel's assumption it is even necessary, as was observed by Green, to suppose that in the absence of all vibrations there is already a certain state of pressure in the medium.

If we adhere to Fresnel's assumption, it is indeed scarcely possible to construct an elastic model of the electromagnetic medium. It may be done, however, if the velocities of the particles in the model arc taken to represent the magnetic force H, which, of course, implies that the vibrations of the particles are parallel to the plane of polarization, and that the magnetic energy is represented by the kinetic energy in the model.

Considering further that, in the case of two bodies connected with each other, there is continuity of H in the electromagnetic system, and continuity of the velocity of the particles in the model, it becomes clear that the representation of H by that velocity must be on the same scale in all substances, so that, if E, *n, I* are the displacements of a particle and *g* a universal constant, we may write Hx =gol, Hy-got, Hz=gat.

By this the magnetic energy per unit of volume becomes zg^2 *(OE)* 2 *(mi)* 2 and since this must be the kinetic energy of the elastic medium, the density of the latter must be taken equal to g^2, so that it must be the same in all substances.

Theories of MacCullagh

A theory of light in which the elastic aether has a uniform density, and in which the vibrations are supposed to be parallel to the plane of polarization, was developed by Franz Ernst Neumann, 2 who gave the first deduction of the formulas for crystalline reflection. Like Fresnel, he was, however, obliged to introduce some illegitimate assumptions and simplifications.

Here again Green indicated a more rigorous treatment. James MacCullagh 3 avoided this complication by simply assumi -ng an expression of the form for the potential energy.

He thus established a theory that is perfectly consistent in itself, and may be said to have foreshadowed the electromagnetic theory as regards the form of the equations for transparent bodies.

Lord Kelvin afterwards interpreted MacCullagh's assumption by supposing the only action which is called forth by a displacement to consist in certain couples acting on the elements of volume and proportional to the components If their rotation from the natural position.

He also showed 4 that this "rotational elasticity" can be produced by certain hidden rotations going on in the medium. We cannot dwell here upon other models that have been proposed, and most of which are of rather limited applicability. A mechanism of a more general kind ought, of course, to be adapted to what is known of the molecular constitution of bodies, and to the highly probable assumption of the perfect permeability for the aether of all ponderable matter, an assumption by which it has been possible to escape from one of the objections raised by Newton.

The possibility of a truly satisfactory model certainly cannot be denied. But it would, in all probability, be extremely complicated. For this reason many physicists rest content, as regards the free aether, with some such general form of the electromagnetic theory as has been sketched.

Bibliography

Antsiferov, V. V. and G. I. Smirnov: *Coherent Radiation Processes in Plasmas*, Cambridge Univ. Press, Cambridge, 1999.

Balakrishan, A.: *Biofuel Crops and Renewable Energy*, Sublime Publications, Delhi, 2011.

Bard, A.J.; Faulkner, L.R.: *Electrochemical Methods: Fundamentals and Applications*, John Wiley & Sons, New York, 2005.

Benenson, W., J. Harris, H. Stoecker, , and H. Lutz: *Handbook of Physics*, Springer, Berlin, 2006.

Bernstein, Jeremy: *Quantum Leaps*, Belknap Press of Harvard University Press, Cambridge, Massachusetts, 2009.

Brauman J. I. and Coulson C. A.: *Supplementary Tables of Molecular Orbital Calculations*, Pergamon Press, Oxford, 1965.

Brush, S.: *The Kind of Motion That We Call Heat*, North-Holland, Amsterdam, 1976.

Callen, H. B.: *Thermodynamics*, Wiley, New York, 1960.

Cox, Brian; Forshaw, Jeff: *The Quantum Universe: Everything That Can Happen Does Happen*, Allen Lane, UK, 2011.

d'Espagnat, B.: *Veiled Reality*, Addison-Wesley, Reading, MA, 1995.

Dahl, P. F.: *Superconductivity: Its Historical Roots and Development from Mercury to the Ceramic Oxides,* American Institute of Physics, 1992.

David Chandler: *Introduction to Modern Statistical Mechanics,* Oxford University Press, Oxford, 1987.

Donald A. McQuarrie: *Statistical Mechanics,* Harper and Row, UK, 1976.

Ehrenfest, P. and T.: *The Conceptual Foundations of the Statistical Approach in Mechanics*, Cornell University Press, Ithaca, NY, 1959.

Freyd, P.: *Functor Theory,* Princeton University, Princeton, New Jersey, 1960.

Gabriel, P. and M.Zisman:*Category of Fractions and Homotopy Theory*, Springer, Berlin, 1967.

Gasiorowicz, S.: *Quantum Physics,* Wiley, New York, 1995.

Gibbs, J.: *Elementary Principles in Statistical Mechanics*, Dover, New York, 1960.

Halliday, D., R. Resnick, and J. Walker: *Fundamentals of Physics,* Wiley, Hoboken, NJ, 2008.

Hoffmann, B.: *The Strange Story of the Quantum*, Pelican, UK, 1963.

Ibach H. and H. Lýÿth: *Solid-State Physics: an Introduction to Principles of Materials Science,* Springer, Berlin; New York, 1996.

Johnstone, P. T.: *Topos Theory*, Academic Press, New York, 1977.

Kresin, V. Z. and S. A. Wolf: *Fundamentals of Superconductivity,* Plenum Press, New York, 1990.

Liboff, R.: *Introductory Quantum Mechanics,* Addison-Wesley, San Francisco, 1998.

Mark W. Zemansky: *Heat and Thermodynamics*, McGraw-Hill, Delhi, 1968.

Mehra, J. and H. Rechenberg: *The Historical Development of Quantum Theory*, Springer-Verlag, New York 1982.

Merzbacher, Eugen: *Quantum Mechanics*, Wiley, John & Sons, Inc., New York, 1998.

Nash, L. K.: *Elements of Statistical Thermodynamics*, Addison-Wesley, Reading, MA, 1974.

Neil Gershenfeld: *The Physics of Information Technology,* Cambridge Univ. Press, Cambridge, 2000.

Planck, M.: *A Survey of Physical Theory*, Methuen & Co., Ltd., London, 1925.

Plymen, R.J. and P. L. Robinson: *Spinors in Hilbert Space*, Cambridge Univ. Press, Cambridge, 1994.

Price, H.: *Time's Arrow and the Archimedean Point*, Oxford University Press, Oxford, 1996.

Prigogine, I.: *From Being to Becoming: Time and Complexity in the Physical Sciences*, W. H. Freeman and Co., San Francisco, 1980.

Rae, A.: *Quantum Physics: Illusion or Reality?*, Cambridge University Press, Cambridge, 1986.

Reichenbach, H.: *The Direction of Time*, University of California Press, Berkeley, 1956.

Squires, E.: *Conscious Mind in the Physical World*, Adam Hilger, Bristol, New York, 1990.

Unruh, W.G.: *Physics Meets Philosophy at the Planck scale*, Cambridge University Press, Cambridge, 2001.

Vidali, G.: *Superconductivity: The Next Revolution?*, Cambridge University Press, New York, NY, USA, 1993.

Wess J. and J. Bagger: *Supersymmetry and Supergravity*, Princeton University Press, Princeton, NJ, 1983.

Index

❑❑❑